Formwork

Formwork

C. J. Wilshere, OBE, BA, BAI, CEng, FICE

Published by Thomas Telford Ltd, Thomas Telford House, 1 Heron Quay, London E14 9XF

First published 1989

British Library Cataloguing in Publication Data
Wilshere, C. J.
Formwork.
1. Concrete. Formwork & moulds. Construction
I. Title
624.1′834

ISBN: 0 7277 1511 9

Typeset in Great Britain by Pentacor Limited, High Wycombe, Bucks.

Printed in Great Britain by
Billing & Sons Ltd, Worcester

Acknowledgements

The Author wishes to thank the following who have kindly granted permission for the reproduction of photographs or diagrams: Council of Forest Industries, Canada, and the Central Electricity Generating Board (Fig. 53), EFCO UK Ltd (Fig. 15), GKN Kwikform Ltd (Figs 10, 12 and 23), John Laing plc (Fig. 2), Laing Technology Group Ltd (Fig. 33), Leada Acrow Ltd (Fig. 16), Mabey & Johnson Ltd (Fig. 46), Outinord St Amand (Fig. 26), Ove Arup and Partners (Fig. 3), Rapid Metal Developments Ltd (Figs 17 and 24), SGB plc (Figs 18, 19, 20, 21 and 25), and the Timber Research and Development Association (Fig. 7).

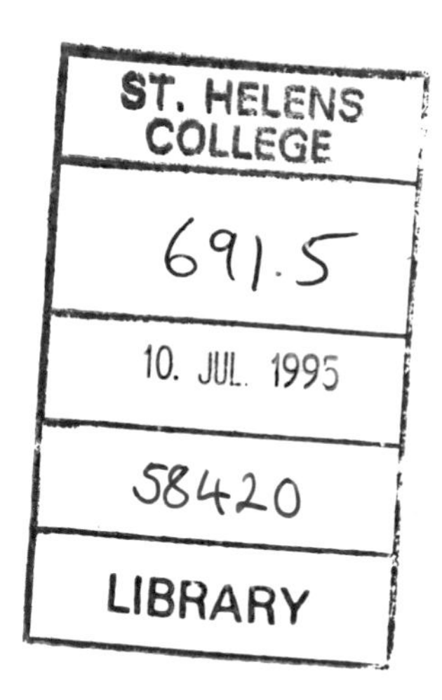

Contents

1 Introduction

Concrete is a material without a shape. For most applications the shape matters, and the concrete has to be moulded or formed. Typically a box will be set up into which it is poured (Fig. 1), and when it has gained strength the box will be dismantled (stripped). An older word used for this box was shuttering (or a shutter, when referring to a single piece). However, attempts have been made to standardise on the other word (originally American), forming (or formwork), which describes both the material and the general subject.

Concrete has been in use for thousands of years. The dome of the Pantheon in Rome is made of lightweight concrete, and the underface shows the moulding effect of formwork used two millennia ago. But it was only at the end of the nineteenth century that the use of concrete became common, with the invention of reinforced concrete.

From being something which the carpenter was left to get on with, formwork has now become the subject of engineering design and is of considerable commercial importance. Both contractors and the suppliers of formwork equipment have a deep interest in the subject and are always seeking to find more economical methods. Developments occur due to the changing ratios of cost between different materials and between materials and labour, changes in the availability of traditional materials, and the introduction of new materials such as plastic. No formwork is typical, but Fig. 2 shows an example.

Formwork is the contact surface and structural backing needed to contain fresh concrete until it can safely support itself. If it is not removed, it is called permanent formwork. If it has to be supported from below, the structural support is called falsework.

The exact distinction between formwork and falsework is not

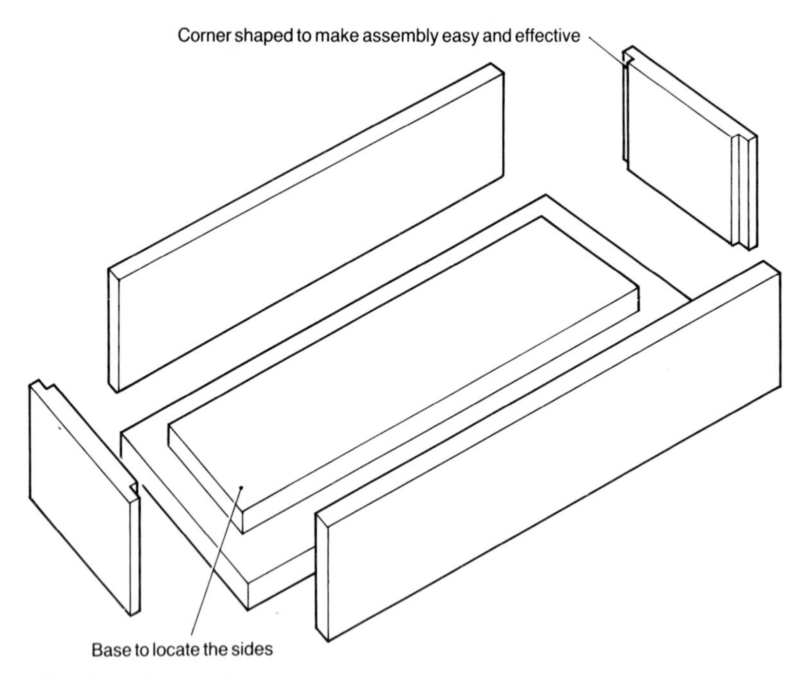

Fig. 1. A box to form concrete

very important, but it is usual to call all material and equipment used to cast walls formwork. This is because the system of loads is self-contained, while for concrete members such as slabs or beams, falsework is used to take the loads to the ground below or to a similar support. For such members, the actual surface material and its immediate support is considered to be formwork. For more information on falsework, reference should be made to the Works Construction Guide[1] on that subject. Other terms will be defined as they arise.

Typically, formwork will be found in the context of a construction contract. The designer, the Engineer, will provide the contractor with details of the concrete work which is to be constructed, but will provide no information on how this is to be achieved. It is the contractor's expertise which the client is buying. The contractor will study the shape of the structure to be created, and decide on appropriate methods to construct it. If the complexity and quantity are both minimal, it may be left to a competent formwork carpenter. However, it is likely today that

Fig. 2. Formwork for columns, walls and slabs

the design will be more scientific. This may be by formal structural design and detailing, so that the personnel on the site have precise guidance on what to do; or it may be that a formwork supplier will be called in and asked to propose methods and to supply the equipment needed. A main contractor may sublet the activity of the formwork to a specialist subcontractor, who would then take the responsibility for this aspect of the work.

The person who actually puts up the formwork will normally be a carpenter or joiner by trade. Many such workers have served a traditional apprenticeship, but today more and more people come into this job by other routes, with less effective training. They may be expected to operate without any guidance, or at the other extreme they may be provided with drawings which set out how to deal with every detail. Some training is available from the Construction Industry Training Board. Courses on design are available from some colleges, but it is a fairly specialised subject and they are few and far between. The City and Guilds Institute conducts examinations on formwork design at a trade level.

Formwork

This guide covers the majority of commonplace formwork applications in the United Kingdom. It is applicable elsewhere, but tradition has a large input into the way formwork is constructed, and some of the ideas here would be considered inappropriate in other countries.

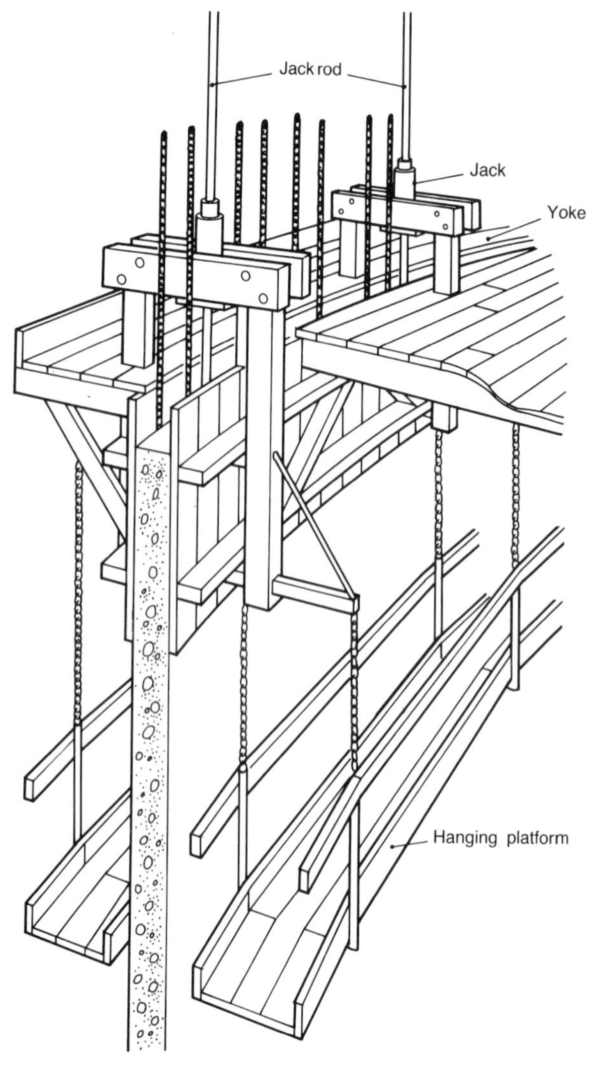

Fig. 3. Slipform

Fig. 4. Simple site precasting of stair flights

Special methods

There are specialised techniques not described in detail in this guide. Two are briefly described below. Refer to *Formwork — a guide to good practice*[2] for further information.

Slipform

Slipform is a method used for tall structures which are constant in cross-section, such as silos or the service cores of buildings. Forms, usually 1¼ m high, are held relative to each other by frames called yokes, and complete plan areas, generally rectangular, are assembled. Jacks fixed to the yokes climbing on rods located within the concrete enable the entire form to be raised as concrete is placed, thus in effect extruding the structure below (Fig. 3). It is possible to vary the layout as the formwork climbs; the process can be used horizontally and on slopes.

Precasting

Instead of casting the concrete in situ, pieces may be made for later transport into position. This differs from general formwork, in that factory-type conditions usually apply. Equipment will be more elaborate, enabling labour saving and allowing good progress to be achieved. Heat may be applied to speed up curing, which limits the choice of forming materials. Fig. 4 shows a simple site example.

2 Requirements for formwork

Fresh concrete is a liquid. It has no natural form, and if uncontrolled becomes a shapeless heap. To make use of it, it is necessary to contain it so that it can be moulded as intended, and it must stay supported in this way until it has sufficient strength to support itself. The formwork is used to contain the fresh concrete, and ensure that it sets in the correct place, is of the correct dimensions of width, height and length, and is of the required flatness. All these dimensions will have required tolerances.

The surface must be acceptable. It is an indication of the integrity of the concrete which is inside and which cannot be seen; if there are voids, doubts will inevitably arise as to the quality of concrete within. It may be that the concrete is to remain visible, and in that case the surface becomes of considerably more importance.

The formwork must withstand safely all the loads to which it is subjected. The concrete will create a pressure which is likely to be increased by the effect of vibration applied to it; on horizontal formwork the concrete may be heaped up temporarily before it is spread. In addition, there will normally be construction loads, of people and of equipment. All these loads must be carried without any failure.

There must be no loss of the constituents of the concrete — or indeed of the concrete itself — either through the joints between the separate pieces of the formwork, or where formwork abuts previously poured concrete. Even absorption by the form face of moisture or loss of some small part of cement will be unacceptable if a high-quality finish is required.

The concrete must not be contaminated. This can happen through the use of an inappropriate release agent, or through too generous application. No pieces of the formwork, for example tape used to seal the joints, or paint inadequately attached to the form face, should get into the concrete. Some species of timber have enough dye in them for this to leach out and tint the concrete. It is important that rubbish is removed from the form before concreting starts, particularly binding wire and nails, which will rust if they are on the surface of the concrete.

It must be possible to dismantle the formwork, and it should be possible to do it without damage either to the concrete or the formwork. It is important that it can be done easily and quickly, as this will also mean that it is being done economically. So that the sheeting will peel readily from the concrete, a release agent is used (see Chapter 5). Almost without exception, formwork should be reuseable. It is to be expected that several uses will be obtained, and that the degree of degradation or deterioration which takes place will be small.

How to achieve these requirements

The concrete to be formed should be assessed and a design chosen appropriate to the skill available. The possible problems, such as leakage, damage to parts which are not strong, and the need for stripping, should be anticipated. Care should be taken to choose appropriate materials and equipment. It is important to check that these are the items which actually arrive on site to be built into the formwork. It should be ensured that the design is followed faithfully during erection. Before concreting, checks should be made on dimensions, on stability, that all fixing has been carried out, that no movements will occur, that the form is carefully cleaned out and that release agent has been used. Stripping should be undertaken carefully, and cleaning and repair carried out for further use.

3 Materials and equipment

The original material used for formwork was timber, and it is still in use today. Other unfabricated materials such as steel are also used. However, the trend is to fabricate materials away from the site to make items of equipment, and as the years go by more and more equipment of varying degrees of elaboration is developed. Despite this, very little material or equipment has fallen out of use, and so the choice gets greater.

Timber

Timber is the most basic material. It is relatively cheap and easy to shape as required. Equally, it is easily cut up and reduced in size, eventually becoming waste. It is easy to fix inserts and the like on the inside of a form face for casting into the concrete; it is relatively light in weight for handling, but it is of limited durability. Surfaces in contact with concrete wear and become damaged, and the framing members can be damaged relatively easily in handling. It has the disadvantage that variations in the moisture content cause significant dimensional changes. The pattern of grain on the face against the concrete tends to leave an imprint, which may be unacceptable.

An introduction to timber for structural use is given in ref. 1, and more detailed information in the BSI Code of Practice for Falsework.[3] Note that timber on site for formwork always has to be classified as wet.

Timber is also the basis of a number of sheet materials. These include plywood, blockboard, chipboard, boards made from wood flakes — oriented strand board and waferboard — and hardboard. All these depend on glue for holding together the timber veneers — the thin sheets peeled from the log — or the particles of which they are made. Long ago plywood was unsuitable for formwork, because the glue which held the veneers together failed in damp site conditions, but nowadays almost all glues are sufficiently durable. Outer veneers should be sanded and without holes — normally referred to as solid. The choice will relate to the quality of the concrete surface required.

For boards which consist of timber particles bonded with a matrix of plastic glue, the low stiffness of this glue leads to a board which is less stiff than plywood or solid timber of the same thickness. The quality of the concrete surface produced from such boards may not be smooth enough, and the deterioration of such surfaces can be fairly rapid. Because the boards are cheaper, however, there are applications where they prove useful.

Plywood. Plywood has some strength in both directions, but because the outer veneers give greater strength in the direction of their grain, the sheet should always span that way. The construction is shown in Fig. 5. Plywood a few millimetres thick is used as a lining material, where it is fully supported from behind. With thicknesses of 12–19 mm the strength of the plywood itself can be used, and the framing members spaced out to economical distances. There are three main types of plywood: Finnish birch; American softwood, including Canadian Douglas fir; and tropical hardwood. Sheets are normally 2439 mm by 1219 mm (8 ft by 4 ft) and, except for Finnish birch, the outer grain runs in the length of the board. The first two types are produced to appropriate national standards, and applications can be based on the data supplied by the producer. Hardwood ply comes from a variety of tropical countries, and obtaining reliable data about them is more difficult.

Blockboard. Strips of timber are made into a panel with veneers

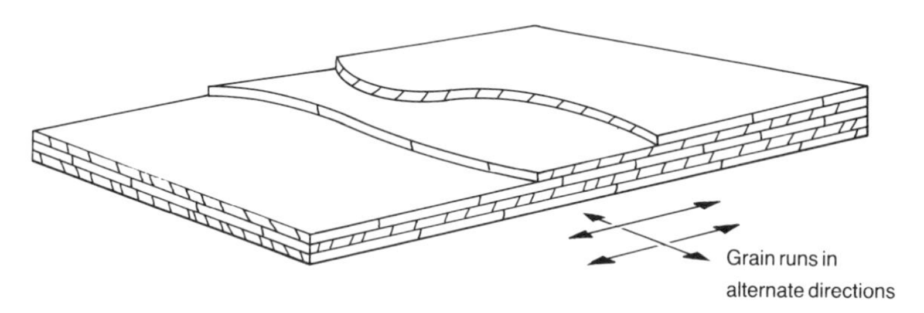

Fig. 5. Plywood construction

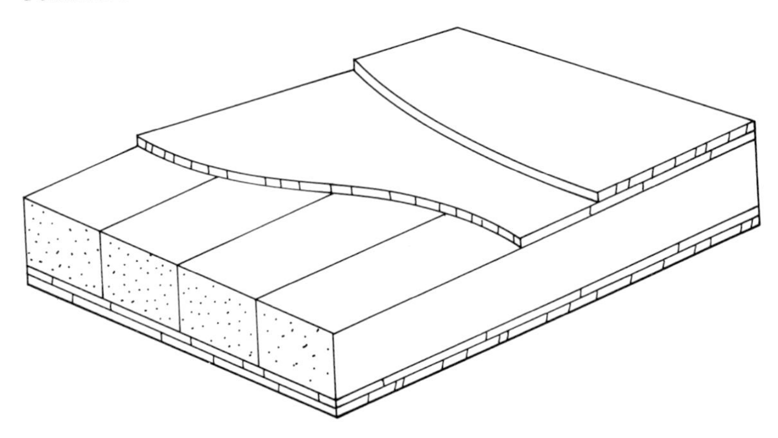

Fig. 6. Blockboard construction

on both faces (Fig. 6). The glue is often inadequate, and unless of high quality, the surface may not be flat enough, as the strips tend to show through.

Chipboard. Small chippings are pressed into sheets. As with all boards depending significantly on plastic as the matrix, the stiffness is poor, and a greater thickness than plywood is frequently necessary. It is heavier than plywood, but for a small number of uses gives a reasonable finish.

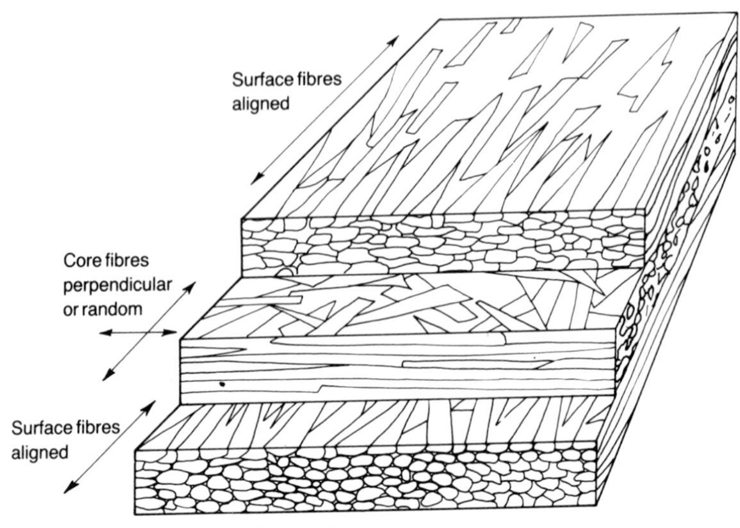

Fig. 7. Oriented strand board

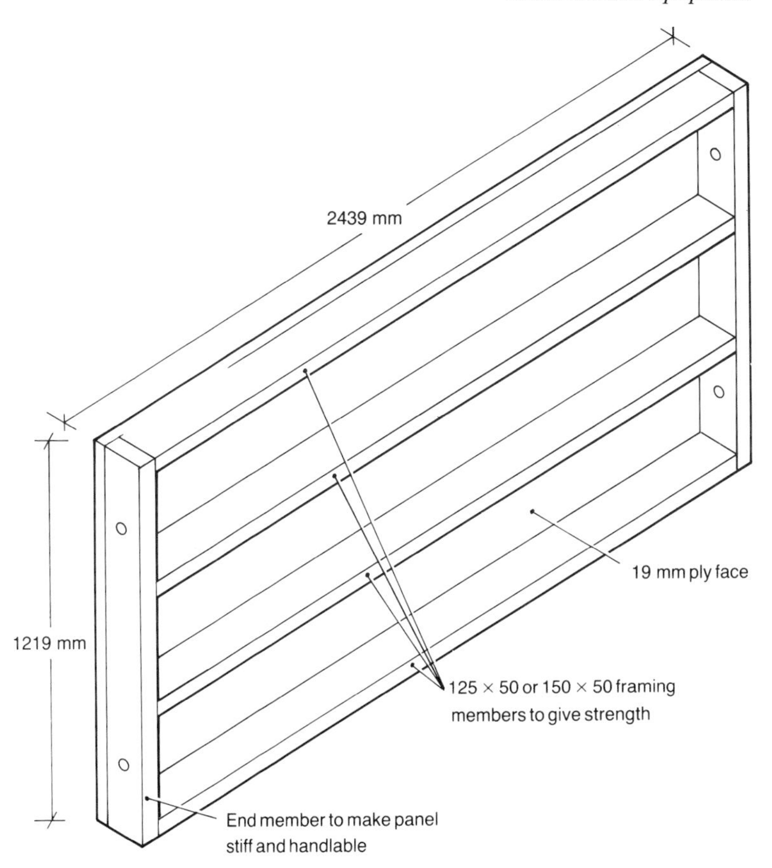

Fig. 8. Typical ply and timber panel

Oriented strand board (OSB). Flat pieces are chipped from solid timber and oriented to give better strength (Fig. 7). The pattern of the chips shows on the finished concrete.

Flakeboard. This is similar to OSB, but the chips are randomly arranged, and it is not as strong.

Hardboard. This is a dense board made from fine wood particles. It is only available in a 2 mm or 3 mm thickness. It is difficult to keep flat due to moisture movements, and is fairly easily damaged, especially at the edges. However, it is useful as a lining for curved work.

Figure 8 shows a typical ply and timber panel; those made from other sheet materials are similar.

Steel

Steel can be used both as sheet material for the face of formwork, and as hot- or cold-formed sections, used to form the support or framing (Fig. 9). It has the advantage that it is durable, although sheet, if too thin, will dent. It is much more successfully fabricated in a shop. It is not easily modified, which can be an advantage. If inserts have to be fixed to the shutter, it will require drilling, and then subsequently plugging for later uses to make the shutter concrete-proof again. It is possible to get magnetic fixings — at a cost! It is a heavy material, and thus the cost of handling it is likely to be greater, as it has to be dismantled into small pieces. Pieces are connected either by nuts and bolts, or often in proprietary systems by some more ingenious time-saving device.

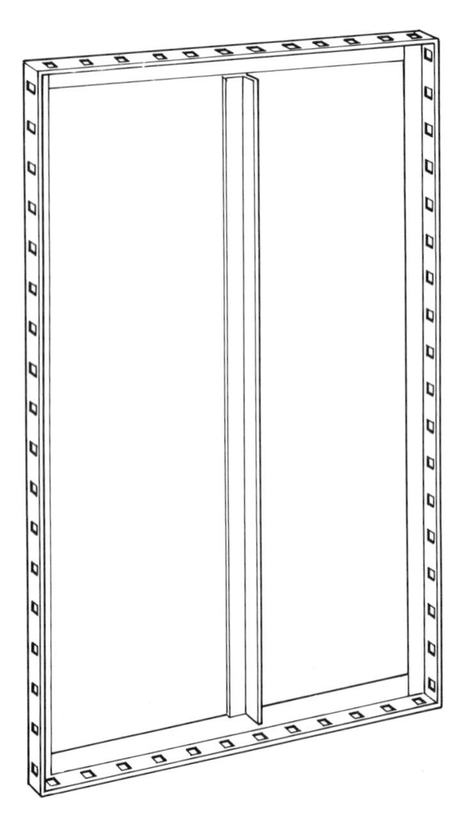

Fig. 9. Steel angles and sheet made into a panel

With both methods, pieces are liable to be lost all too easily.

Steel is dimensionally stable, but is liable to rust. Rust will come off on the concrete, even if it is not so severe as to corrode the form face. Appropriate release agents should prevent this problem (see Chapter 5 on finish). Steel is available in various grades of strength, but with formwork the usual criterion for design is the stiffness of the material. Because the stiffness of all grades of steel is the same, there is seldom any advantage in spending more money for a higher grade of steel. If any welding has to be done, it is best to use lower grade steel to make the welding easy and successful. Because it is a hard unyielding material, demountable joints between adjacent pieces of steel are difficult to make in a watertight fashion, to proof them against leakage.

Aluminium

The other metal which is now coming into common use is aluminium. It has long been a problem that the reaction between concrete and aluminium results in corrosion of the aluminium when used as a face material, and, when used in this way, a suitable release agent is vital. Aluminium used to be expensive, but is becoming more competitive. While it is not often used as a basic material, the use of special extrusions as beams is popular (Fig. 10). Working in conjunction with the slots in the extrusions,

Fig. 10. Extruded aluminium beams

Fig. 11. Plastic used to make a circular column form with a splayed top

devices to make attachment of plywood and other members simple and quick are useful (see proprietary equipment, p. 16).

Plastic

There are always hopes expressed that plastic will take over the majority of formwork applications, but it has two significant disadvantages which have so far prevented this happening: the elastic modulus or stiffness of all plastic is relatively low, thus requiring a considerable thickness of material to achieve adequate resistance to deflection; and the price is dependent on the price of oil, which in the long term is likely to rise. Nonetheless, there are useful applications for certain plastics. Because of the stiffness aspect, these are mostly for surfaces which are not flat.

For example, glass-reinforced plastic (GRP) makes a satisfactory circular column form (Fig. 11). If suitable stiffening framing is added, it can also be used successfully for rectangular columns. GRP is also useful for three-dimensional curves, such as column heads, or formers to make a waffle floor (Fig. 12). Materials which soften on heating and then stiffen again on becoming cool (thermoplastics) can be shaped for similar purposes. Making connections to pieces of plastic like these is difficult, and so appropriate uses are applications such as waffle formers, or as linings in forms for creating patterned finishes where separate pieces are appropriate. Plain thin plastic sheeting as a liner to a form is not satisfactory, because it is extremely difficult to fix the ends successfully and because despite the apparent flatness, almost inevitably the edges are longer than the centre and the plastic cannot be made to lie completely flat. Thermal movement is frequently a problem with all plastics. However, plastic is used successfully as a surfacing material bonded in manufacture to plywood sheets. Melamine and phenol formaldehyde are often used, but different suppliers offer quite a variety of materials. In general, the thicker materials are better but dearer.

Expanded plastic is of considerable use, and blocks of expanded polystyrene are common on any construction job where recesses and pockets have to be formed in the concrete. It is useful for void

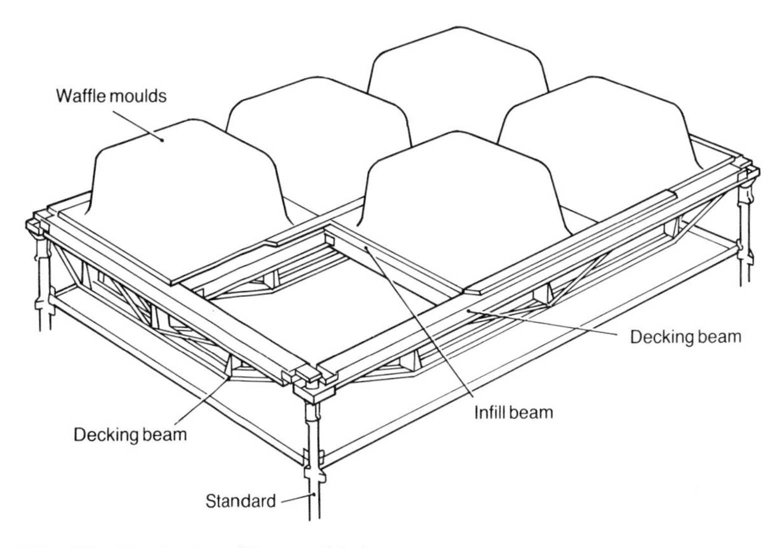

Fig. 12. Typical waffle moulds in use

forming in bridge decks and the like. There is now available a moulded commercial shutter panel made of heavy duty expanded plastic.

Use of materials for permanent formwork

In some cases it may be possible or desirable to leave the formwork in contact with the concrete. For example, a brick or block wall can be used to form a pile cap; various materials can be used to form the soffit of a bridge deck between precast beams (Fig. 13). GRP and glass reinforced cement are appropriate.

Proprietary equipment

Panels made of plywood and timber framing are frequently constructed on site; they are not available as commercial items. However, steel panels are available, and come in two basic types. In the traditional type the sheet of metal has four edge angle members and, depending on the size, some intermediate framing. The panel is used with separate framing members to provide the

Fig. 13. Commercial sheeting used as permanent formwork on precast beams

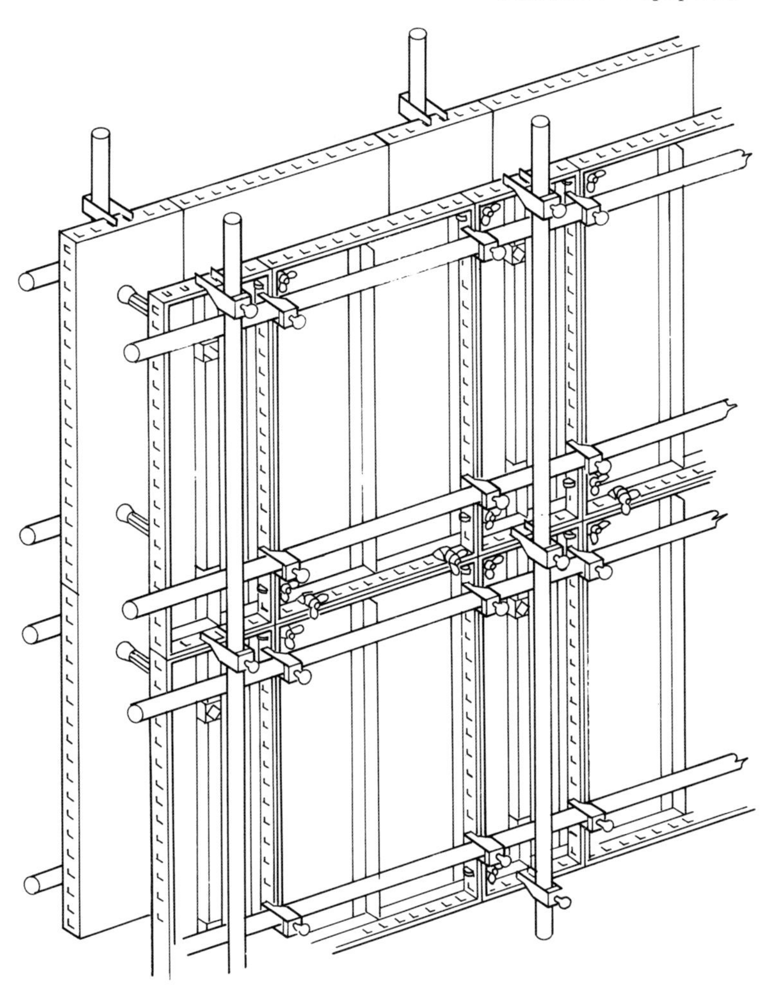

Fig. 14. Steel formwork assembled ready for use

basic strength (Figs 9 and 14). Alternatively, a heavier panel can be used, in which the support system of ties (see below) from one face to another can be put directly through the panel, and, where additional framing is used, it is merely there to ensure that the structure is straight. Heavy panels of this type can also be assembled with panels at the sides and underneath to make a complete beam form which only requires a support at each end (Fig. 15).

Fig. 15. Heavy steel formwork supporting itself between columns

Because of the advantages of plywood — facility for fixing and weight — a number of panels are available which are constructed as a steel frame with a plywood face (Fig. 16). Normally the frame would come to the face of the concrete, so that the metalwork protects the edge of the plywood, but it is also possible to obtain panels in which the framing is totally behind the plywood. The practical difference is in the appearance of the concrete. Where the steel frame encloses the ply, at a panel joint not one line but three are seen, thereby emphasising the joint line considerably. Where ply meets ply the joint will be much less conspicuous when the panels are new, but damage to the ply is likely to occur more rapidly, leading to ragged joints.

Metal-edged panels can also be made of aluminium. This gives the advantage of further reduction in weight, but at a higher capital cost, and with the risk of the concrete sticking to the lines of aluminium which are in contact with the concrete. The effective use of release agent is then more important.

Systems

All these panels normally form the basis of 'systems'. This term describes a group of components which will provide a complete

Fig. 16. Ply-faced metal-framed formwork

solution for simple problems, and with little additional material cope with more difficult cases. Individual components can usually be used in other contexts.

Walings

As an alternative to assembling panels, it is practical to assemble a number of sheets of plywood and provide framing members, called wales or studs, behind them. Wales are horizontal members, and studs vertical. The proprietary framing members available today are of steel or aluminium. The steel member is a channel section with holes through which fixings can be put to attach the plywood, for example short large-diameter wood screws. The equivalent aluminium wale is more elaborate, making full use of

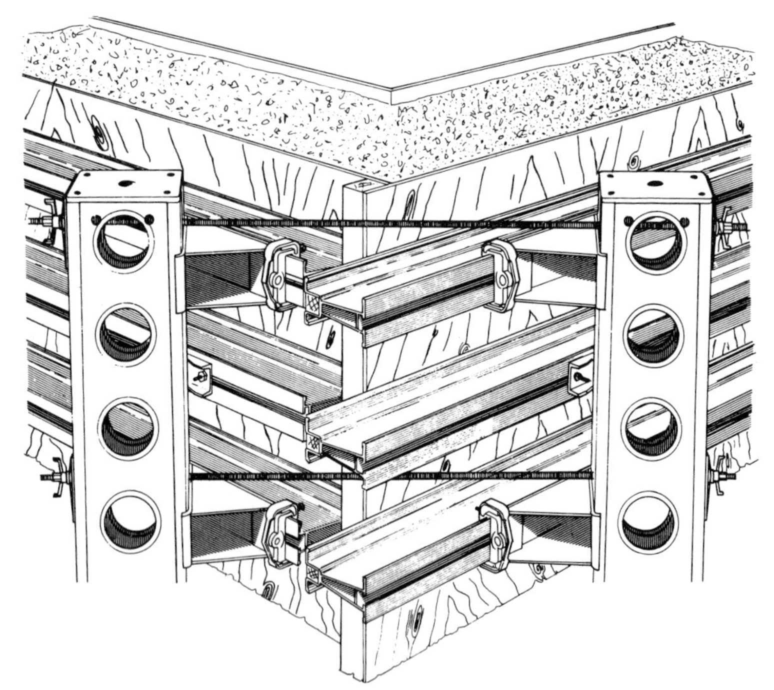

Fig. 17. Use of aluminium walings and accessories

the possibilities of extrusion. Typically, on one face there will be a recess in which a piece of timber is made captive, enabling ply to be nailed or screwed. On the opposite face there will be a slot into which a T head bolt can be held captive, enabling the outer framing to be attached with a minimum of difficulty. Thus all the awkward problems of assembly are dealt with by this one unit. All systems offer fittings which enable connection to be done simply and quickly, and enabling the various problems such as corners to be solved easily (Figs 10 and 17), but such fittings are expensive. The proprietary metal waling member is almost always lighter and stronger than the equivalent timber member, and the design will utilise this extra strength by spacing the outer (primary) framing members further apart.

Soldiers

Vertical members forming the main external framing are called soldiers (or wales if used horizontally). The normal construction is of cold-formed sheet steel, typically 250 mm deep, making units which are similar to a pair of rolled steel channels back to back (Fig. 18). The extra complexity gives them a better strength-to-weight ratio, and enables fixings to them to be made more easily. Since these were first available the strongest soldiers have doubled in strength, which enables a wider spacing to be used, with consequent economy. Most soldiers have fittings available which, for example, provide a jacking system outside the base to facilitate plumbing. Those available in other countries include framed soldiers, which are much deeper than the cold-rolled sections. Fabricated timber sections are also available elsewhere, either of laminated timber forming an I section or as a truss.

Equipment for curved formwork

To form curved surfaces, there are a number of items available. Steel shutter panels are available without end framing, so that a curve in one direction can be made. This is normally used in conjunction with a scaffold tube which has been rolled to the appropriate radius. Where it is intended to use plywood, there are systems involving the attachment of parallel framing members to which are attached screw mechanisms enabling the ply to be pulled to the appropriate radius (Fig. 19). For circular columns, forms can be made specially in steel or plastic, and it may be possible to acquire these secondhand.

Ties

Formwork to vertical members such as walls normally requires ties to resist the pressure of the concrete. These ties connect the outer framing members of one panel with those of the opposite side of the wall so as to balance the forces. A tie must hold each pair of forms in the correct position, neither creating an hourglass wall nor allowing it to bulge, and it must continue to do so when the concrete is placed. Ties are of two types: those which are withdrawn totally after use, and those in which the centre portion remains permanently in the concrete. In the former case the problem of waterproofing may arise, although there are acceptable methods of waterproofing a hole. It is normal to require any left-in steel pieces to be at least as far from the face as the specified reinforcement cover, to minimise the risk of rusting.

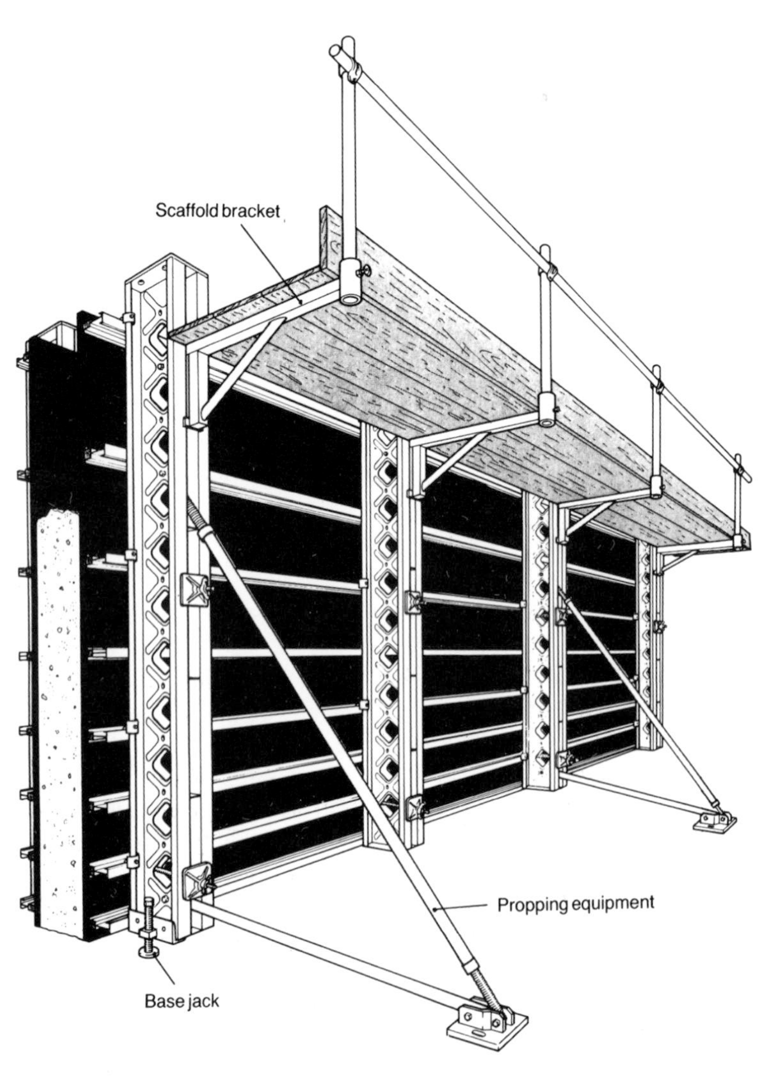

Fig. 18. Typical soldier and accessories used with aluminium walings and a plywood face

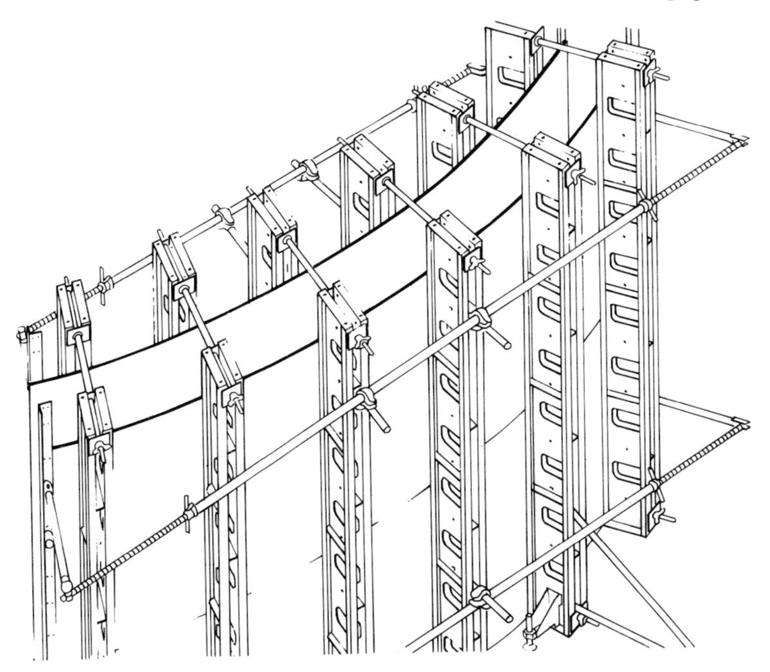

Fig. 19. One of the systems for curved formwork

Extraction of ties can be achieved either through the use of a sleeve to prevent the concrete touching them, or by using ties of tapered section, so that they can be extracted in one direction. Most proprietary versions of the plain tie are based on a Dividag bar (Fig. 20), and many manufacturers provide appropriate nuts and washers to transfer the loads. Plastic tubing is used as the sleeve, frequently in conjunction with special plastic end-pieces which enable a reasonable seal to be achieved with the formwork, thus limiting concrete leakage. The plastic ends enable the tie to be done up tightly, ensuring that the width of the wall is correct. Potential problems with this type of tie are the ingress of grout, and difficulties in extraction if the bar has become bent.

The most common type of leave-in tie in use is the she-bolt (Fig. 21). A threaded rod of appropriate length, located in the centre of the wall, engages each side with a bolt, in the end of which a hole has been drilled and tapped. The portion of the bolt which goes into the concrete is tapered, and at its outer end it is threaded

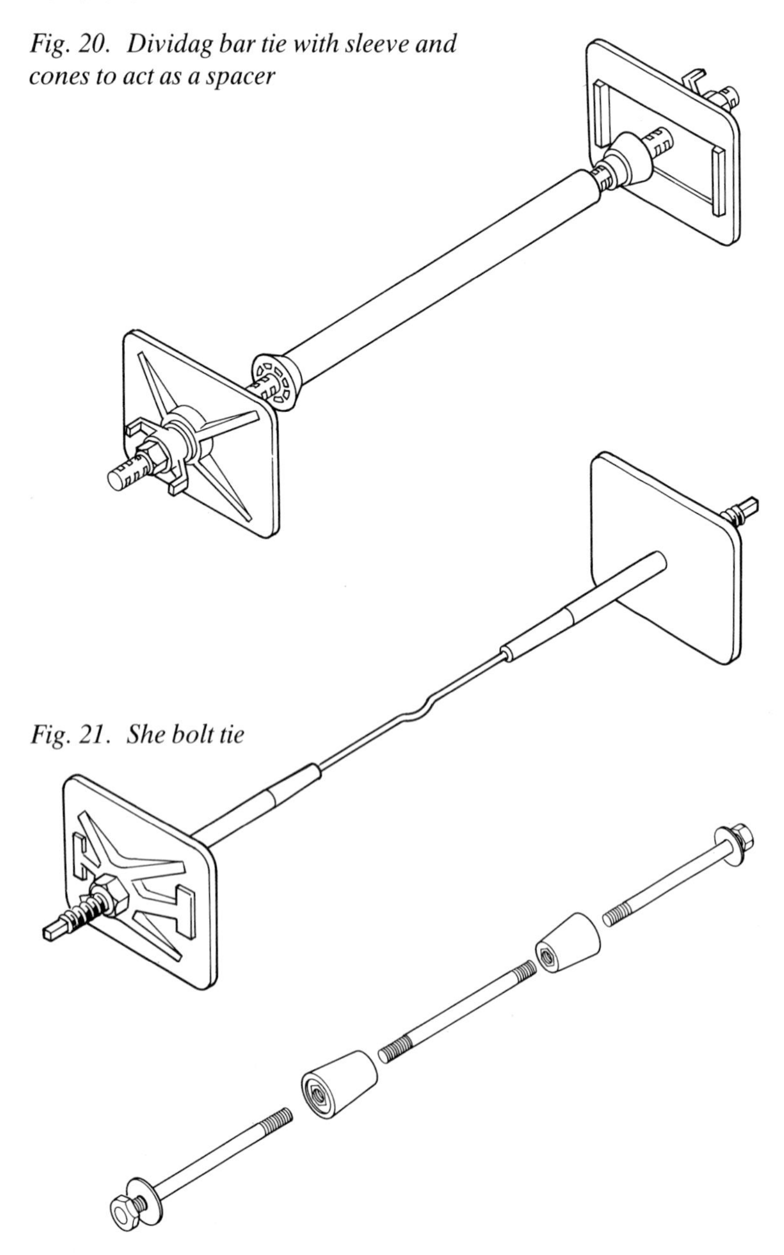

Fig. 20. Dividag bar tie with sleeve and cones to act as a spacer

Fig. 21. She bolt tie

Fig. 22. Rod and cone tie

externally to engage with an appropriate nut and washer to transmit the loads from the framing of the formwork. This is a successful system, but has some potential problems. There is no method of spacing the two shutter faces apart, and so separate spacers are usually essential. It is difficult to be certain that the rod is exactly in the centre of the wall; if it is not, the cover that is required may not be obtained. When attempting to strip, the she bolts may stick; the rod in the concrete, despite being provided with a crimp, may turn, thereby ruining any waterproofing which would have otherwise have been achieved.

Another type of tie is the rod and cone (Fig. 22). A bar with threads at both ends is in the centre of the wall, and is screwed at each end into cone units each comprising a threaded piece of metal within a plastic cone. At the outer face of the cone which abuts the formwork face there will be further thread to enable a bolt to be attached, and the facility for a box spanner to unscrew the cone from the rod in the wall. Conventional bolts would normally be used with this system. There is no difficulty in spacing the forms if the length of the tie unit itself has been determined correctly, because there are pins through the centre of the cones which enable the tie and twin cone assembly to be tightened to an exact length. There is seldom any difficulty in extracting the cones. Where waterproofing is important, the detail at the centre of the tie unit may be specially arranged; typically, there would be a crimp. For waterproofing purposes it is possible to weld a plate on, so that any leakage water has a much more devious path to follow and the risk of the tie turning is eliminated. Some systems offer a rubber washer which grips the rod tightly.

There are occasions when it is necessary to fix formwork to previously poured concrete, but it is not practical to have a tie through to the opposite face. All systems offer components for this purpose, which are known as anchors. These will be cast in near the top of the previous lift, so that the shutter to be erected above can be tied to them in due course. Their strength is dependent on the strength of the concrete.

Soffit formwork

The underside of a ceiling or lintel is called a soffit. If it is to be made of concrete, soffit formwork will be needed. Where the form surface is of plywood or similar sheeting material the layer of support immediately underneath may be of timber, or the framing members described above. Primary members below these upper layers enable the vertical loads to be supported at a limited number of points. This may be by props or scaffolding such as tube and fittings, or the proprietary type.[1] An alternative approach to the support of sheet members is the skeletal system. This consists of a series of vertical supports (props or scaffolding), between which span a series of primary beams. These are provided with a shelf or recess along each edge, such that other beams can span between pairs, thus filling in the area with a series of supports sufficient to limit the span of the sheeting. The basic concept for such a system requires that it be stripped, starting with removal of

Fig. 23. A skeletal system being assembled

the basic support. It must therefore remain in position until the slab is quite capable of supporting itself. There is a variant, in which the top of the support is a small separate piece, typically 150 mm square, which forms the soffit at the point immediately above the propping point, and which is arranged to stay in position while all the beams and the sheeting material above them are lowered a small amount and removed. A skeletal system is a practical way of supporting trough or waffle formers for ribbed or coffer floors. It is called a 'quick strip' system (Figs 12 and 23).

As an alternative to sheet material, panels may be used in conjunction with the skeletal system described above. Some systems have beams with pins; the pins locate in the underside of the panels and with the tops of the supports to prevent lateral movement and ensure correct positioning.

Where it is desired to span greater distances and keep the space below the soffit substantially clear, telescopic centres may be used. A centre has two sections, each with a lug on one end, which telescope one into the other, and which can be set spanning from one wall to another up to a distance of 5 m (Fig. 24). They are available in different sizes and strengths. See Chapter 6 for practical details.

For heavier and longer spans there are other items of equipment available, and it is possible to use Bailey bridging equipment for this purpose.

Larger components

All the equipment described above has to be dismantled into relatively small pieces to be carried forward and used again. Where there is repetition, and no difficulties in handling such an item, the soffit formwork can be made into a large panel and be

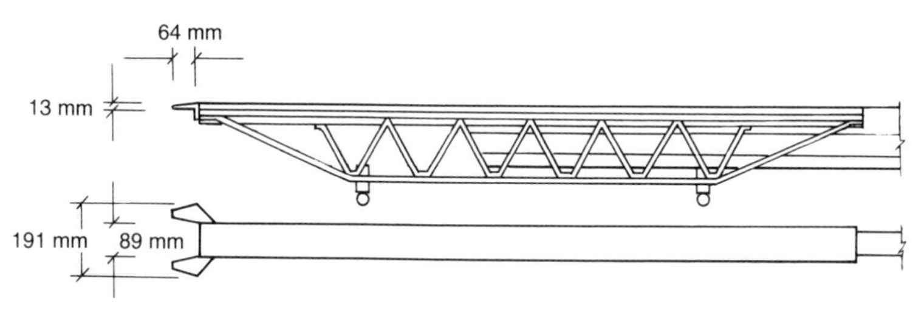

Fig. 24. Typical telescopic centre

Fig. 25. A table form being extracted by crane

handled either with or without the support framework underneath. It may be handled forward on a level base, for example in a reservoir; on a multi-storey building the crane will be used to extract it like a drawer and place it in the next bay to be concreted. Such equipment is normally provided as a framework, on which plywood is used for the surface. The framework is usually constructed from a kit of steel or aluminium parts, and can be made up to approximate closely to the application in hand. Units up to 30 m long can be handled by crane in appropriate circumstances. Small units are uneconomical. Units are known as tables, trollies and flying forms; an example of a table form is shown in Fig. 25.

Also available are apartment forms. These are sometimes

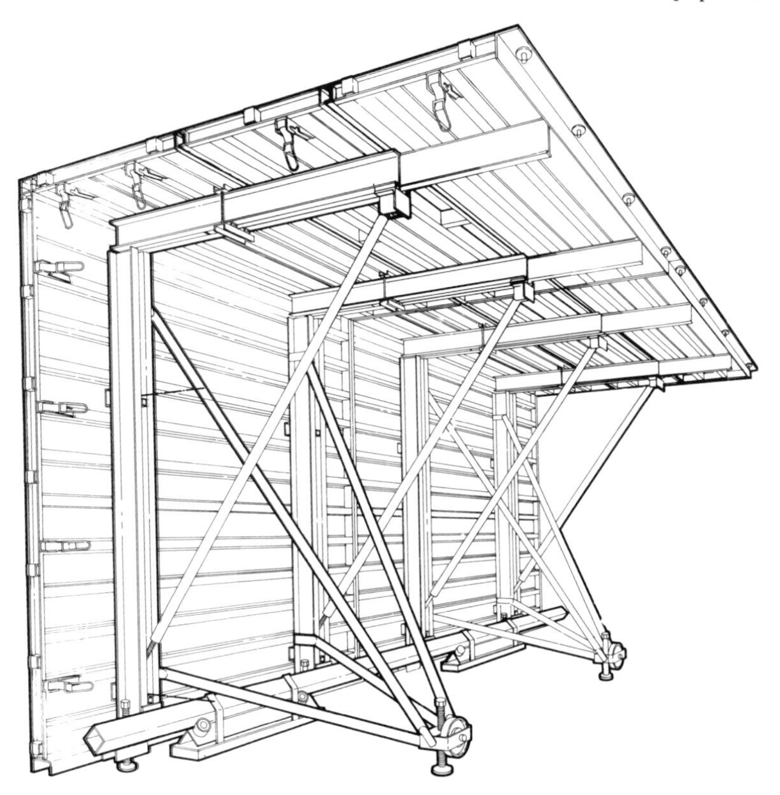

Fig. 26. A half apartment form

referred to as tunnel forms, because they have forming surfaces both across the top and down the sides or ends. They are made both as a complete inverted 'U' and in half units (Fig. 26). They are used in the construction of multi-storey buildings where wall and soffit are poured at the same time. Special release mechanisms are provided to enable stripping to take place. To ensure a rapid re-use of this relatively expensive equipment, it is common to provide heating for this operation, in some cases built into the formwork itself.

For situations with extensive repetition, special travelling shutters are made, which may be hydraulically operated. An example is for tunnel lining, where forms are designed to telescope so that one will pass through another in front, ready for re-use.

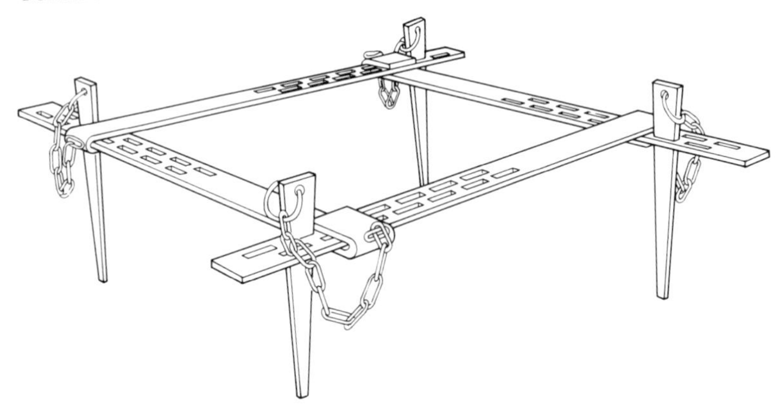

Fig. 27. A standard column clamp

Clamps

Two types of clamp are commonly used. The column clamp has been in use for many decades, and comprises four flat straps of metal connected at the four corners with wedges; it enables

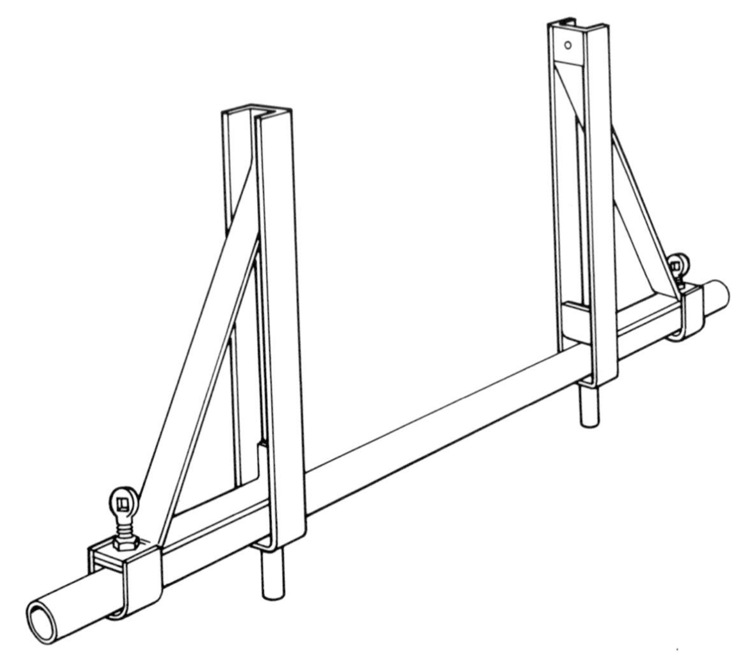

Fig. 28. An adjustable beam clamp

column forms to be restrained simply and economically (Fig. 27). For larger cases, other more substantial equipment is needed. The beam clamp comes in a number of designs. Essentially it is a horizontal cross-bar which goes below the beam, on each end of which is fixed a bracket (Fig. 28). These two brackets or arms are pushed tight up against the side panels and locked, providing support during the concreting operation. Such equipment is used efficiently when both brackets and bar underneath can be withdrawn the next day, but some are designed to be part of the soffit support system, and thus cannot be completely stripped until the whole soffit is dismantled.

4 Design

Using the widest interpretation of the word design, all aspects leading to the actual detailing of the formwork can be included, covering both practical arrangement and structural design. This chapter deals with the latter part.

Formwork has always been designed by the elastic or permissible stress method; data on the construction side is not available to use the alternative — the limit stage, gamma factor or statistical method. It is practical to take conventional loadings, and assume that these are equal to characteristic loadings, but there is no established body of opinion on appropriate gamma factors to use. The nature of formwork and the variety of contexts indicate that a partial factor approach could be of considerable use. For example, there are varying levels of workmanship and supervision, different qualities of material and equipment, and differing levels of checking involved. Pending authoritative resolution of these uncertainties, the permissible stress method should continue to be used.

Factor of safety

It can be argued that the factor of safety should be increased to cover the comparatively casual attitude of operatives and supervisors to temporary works, and to allow for the fact that reused materials will be built into the structure. On the other hand, it could be argued it should be decreased, because the top limit of loading is known with much more certainty. Allied with this is the short-term one owner situation, which eliminates an area of doubt which arises in permanent construction. These arguments have been used to adjust the safety factor from that normally used, but it is unusual to do so.

Stiffness

While strength is important, it is frequently stiffness which is the controlling factor in the design. This is of direct importance to the flatness or straightness of the concrete. From the aesthetic point of view, if lack of flatness can be seen the concrete would be considered unacceptable. So long as light comes from the front, rather than tangentially, there will be little problem, but where the sun can glance along the face of the concrete it will show up every ripple in the surface. On the practical side, it depends what will be put against the concrete; a window frame must go in the space intended; and partitions must go under soffits to create rooms. Where none of these points is of consequence, it may still be desirable to limit deflection, both because extra weight and cost of concrete could be involved, and because during actual placing of concrete, movement of that already consolidated may cause it to crack. Occasionally a requirement that the concrete shall be flat without any tolerance is made, but the generally accepted figure is $\frac{1}{270}$ of the distance between the immediate framing members supporting the facing. Where a specific tolerance is given, it is necessary to add the deflection of the various components in the structural system. For camber, see Chapter 6.

Another aspect of the stiffness is that at construction joints. Where the shutter has to nip on to existing concrete and provide a seal against leakage, the stiffness required is likely to be considerably greater.

Concept

It should be emphasised that the load path should be carefully considered to make sure that it is effective and rational. Unless each member spans a greater distance than the one nearer the concrete, whether such a member should be used in the design at all should be carefully considered. For example, when designing a wall form with studs the following common sequence should be considered (Fig. 29).

(a) The pressure exerted by the concrete must be resisted by the facing material which spans between studs.

(b) The spacing of the studs will be dependent on the strength of the facing material used.

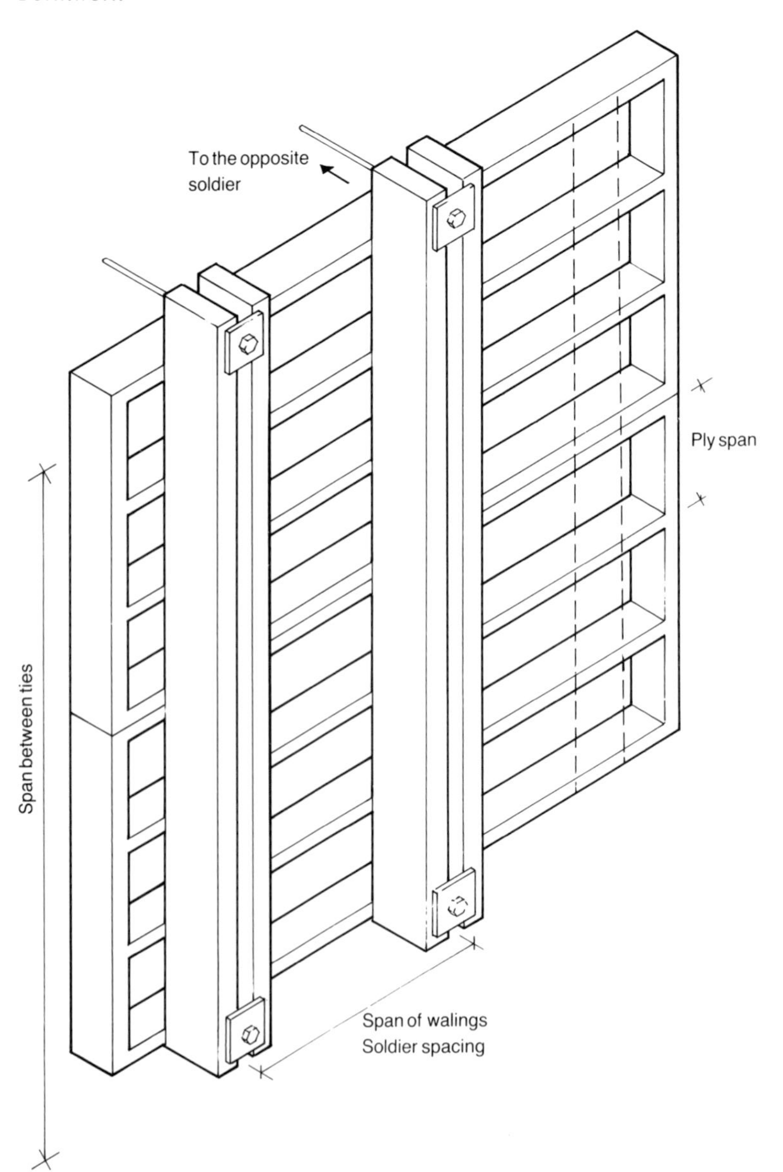

Fig. 29. A typical structural arrangement of formwork

(*c*) The studs will span between walings.
(*d*) The studs must be designed for the loads imposed on them by the facing material.
(*e*) The spacing of the walings should be chosen to suit studs of a convenient size.
(*f*) The walings will span between ties which go through to the opposite face.
(*g*) The walings must be designed for the loads imposed on them by the studs.
(*h*) The spacing of the ties should be chosen to enable walings of a convenient size to be used.
(*i*) The load in the ties will be determined from this. The most economical design will result when spacing between studs, wales and ties increase progressively.

Elaboration

Apart from vertical loads, which can be predicted fairly accurately, other loads can vary considerably. For this reason, and those discussed below, there is little point in adopting elaborate methods of calculation when frequently the intentions of the designer will not be carried out precisely on site. The details which present potential problems are

(*a*) non-continuity — short pieces of material where a single continuous one had been anticipated
(*b*) minor 'improvement', where the site workers make small changes
(*c*) whether or not moment transmission or fixity occurs
(*d*) tolerances achieved on site.

It is necessary to consider the worst or most onerous situation likely to arise, and not to specify an arrangement which could inadvertently lead to excessive deflection or stresses.

Where crane handling is to be used, the design should provide for attachment points to the form with a secure load path to the body of the form.

Loads

The principal loads to be considered are concrete, construction activity, and wind on vertical members. The concrete produces both vertical and horizontal loads. The weight acts vertically. In general it has a specific gravity of 2·5, but concrete of other densities is used. It is not normally necessary to make any adjustment for the amount of reinforcement, but if this is very dense it should be specially calculated.

The horizontal pressure from normal concrete can be considered at its simplest as that produced by a fluid 2·5 times the density of water. However, for taller pours and for slower pours the effects of the internal friction and the setting of the concrete in some cases reduce this pressure considerably. The faster the rate of pour (measured as metres per hour of filling the form vertically) the greater the pressure. This results in columns of large cross-section having the greatest loading, but very small ones do not reach anything like the predicted figures. Details of the most recent British research are given in CIRIA report 108.[4]

An allowance must be made for the weight of men and equipment. The minimum figure to be taken is 1·5 kN/m^2, but consideration also must be given to the materials which may be stacked before the formwork is removed. Both mechanical plant and concrete pumping can cause horizontal loads as well as vertical. Provided the heaping of concrete as it is deposited from a skip is kept well under control, there is no need to allow specifically for extra loading, as the load is of short duration and can be contained within the margin of strength of the formwork. Where materials are to be placed on the concrete, either these must be carefully assessed, or a substantial figure allowed to permit unsupervised stacking. Reinforcement is much heavier than it appears at a casual glance. The guard rail at the edge should have a lateral capacity of 0·75 kN/m, but a figure of 5 kN/m is needed where the public have access.

Environmental loads

The main factor to consider is the wind. The BSI Code relating to wind loads[5] gives the detailed information which is needed for calculations. These loads are particularly significant where wall formwork is erected, as large areas are very similar to sails on a boat.

The basic wind speed S is easily established from the map.[5] It is usual to take S_1 as equal to 1·0, although consideration should be given to the 1988 amendment of the wind Code[5] on this point. For S_2, which covers the height above ground and the type of exposure, reference must also be made to this Code.[5] The statistical factor S_3 is normally taken to be 0·77, covering a period of 2 years. Should it be anticipated that the structure will be up for longer for any reason, it should be increased upwards. The new factor S_4 may enable a small decrease to be adopted, depending on the particular orientation of the formwork, but is unlikely to be of importance, and so can be taken as 1·0. The actual design pressure can then be obtained from the table in the Code. Similar data has been extracted and is available in the falsework Code,[3] and a more basic approach is given in the formwork guide.[2]

The conversion of wind pressure to actual load depends on the area of the formwork and any surrounding parts, its shielding, and the shape factor of the various pieces.

There are a number of other environmental loads which should be considered. These include water, where the support to the formwork may be a river. As well as the direct effect of flowing water, there may be debris and waves. Design for snow is seldom needed, as normally this will only be a load when the constructional load is absent and thus is not additive. Similar remarks apply to ice, but under adverse weather conditions there can be a substantial build-up.

Where formwork is supported on a temporary vertical structure, it is important to have adequate lateral ability. It is difficult to quantify all the forces likely to lead to instability, but some have been discussed above. It is normal to consider 2·5% of the vertical load as a notional horizontal force, on the basis that the vertical members are unlikely to be truly vertical. For the purposes of design the following should be considered

(*a*) wind
(*b*) erection tolerances — at least 1% of the vertical load
(*c*) out of vertical by design
(*d*) concrete pressure
(*e*) water
(*f*) waves
(*g*) dynamic and impact forces from mechanical plant
(*h*) movement due to prestressing cables.

In no case should a figure less than 2½% be used.

Calculation

Actual design follows normal structural practice. Reference should be made to appropriate design codes. In particular, see the formwork guide[2] and the Code of Practice for Falsework.[3]

Standard data

The falsework Code[3] and the booklet 'Safety in Falsework for in-situ beams and slabs'[6] provide information on standard solutions, which will thus obviate the need for calculations. It is important to check that all the limitations given are followed.

5 Concrete finish from formwork

The finish or outer surface of the concrete is all that we normally see. It is a measure of the integrity of the concrete, and we judge whether or not it is satisfactory from this appearance. In some cases it is also necessary for the surface of the concrete to be visually acceptable. Just what is acceptable is often a matter of subjective judgement, but in general a smooth, uniformly coloured surface is desired.

The finish is a function both of the concrete and of the formwork in which it is placed. For the concrete, the main aspects are the mix design, the variability of the concrete which is used, and the way it is placed. These are outside the scope of this guide. As far as the formwork is concerned, the main items are the surface quality of the form, the joints, and the conduct of the whole formwork operation.

Structural concrete

The judgement of concrete integrity is made by examining the following aspects.

Blowholes. Small holes in the face of the concrete up to 15 mm across are called blowholes, and are caused by failure to get rid of the air which arrived with the concrete.

Leakage. The dark marks associated with joints are normally caused by the mix losing water, changing it so that when set it is of a darker colour. This is called hydration discolouration. In more extreme cases, there will be loss of fines too.

Honeycombing. This is normally a case of poor compaction, where there are holes in among the larger aggregate which have not been filled with sand and cement. It can also be caused by the mortar part of the mix escaping from open joints in the formwork.

Contamination. The items likely to cause concern are the effects of the release agent; leaching of dye or even splinters from materials such as timber; the retarding effect of sugar in the timber; and the appearance of rubbish in the face, for example binding wire or nails at the lift line.

All of these items may give cause for concern, although in most cases the consequences will be of minor importance. Examples are shown in Figs 30–32.

Fair-faced concrete

For fair-faced or visually acceptable concrete all the items above have a greater relevance, and there are a number of others.

Colour variation. Release agents can be the cause of discolouration (oil discolouration); there can be differences in the way different parts are cured; the absorbency of different parts of the formwork may be different; and glassy smooth impervious surfaces can give rise to a blackening of the concrete.

Fig. 30. Poor concrete finish showing leakage marks along form joint lines and blow holes

Fig. 31. The bottom edge of a beam showing honeycombing and the entrapment of some paint film

Fig. 32. Concrete which has been lightly grit blasted; rubbish has been exposed on the joint line and the blow holes are more conspicuous

Pattern. Patterns created by the pieces of sheeting material, ties and construction joints are almost impossible to hide. Initial colour differences between successive pours often persist.

Contamination. Form scabbing is a problem which occurs when formwork has whiskery pieces on the surface which get caught by the concrete and pulled from the form. Both tape put over joints and paint applied to the surfaces of forms may become partially detached during concreting and similarly be trapped in it.

Crazing. Very smooth form faces give rise to very smooth concrete which has a much higher tendency towards crazing — a fine network of cracks.

Lack of flatness. The design of the shutter with an excessive span of the facing across its supports will give rise to rippling; physical damage — minor indents — will produce corresponding

Fig. 33. Markings due to excessive mould oil

lumps on the concrete surface. At joints, failure to align adjacent panels will result in a step in the concrete. Where the design or execution are inadequately thought out, shutter movement may result.

Damage. This is caused by stripping too early, or perhaps by passing mobile equipment knocking the concrete.

Sympathetic vibration. Occasionally the frequency of the vibrator and the design of the shutter are attuned, so that the concrete mix varies depending on whether it is opposite the framing member of the panel or the much more mobile centre of a panel. The result is a difference of appearance, but unfortunately it is not known how to predict when it will happen.

Examples of some of these faults are shown in Figs 30–36.

Fig. 34. Concrete showing the pattern of ties and formwork

Fig. 35. Small slivers of wood caught in concrete

Fig. 36. Rippling of concrete — formwork not stiff enough

Controllable aspects

Something can be done about all the problems listed above, but it may be that the effort needed to achieve success is disproportionately large. Placing and compaction techniques are largely outside the scope of this guide, but the following sections discuss the main areas where action can be taken.

Design aspects

Forms should be designed to be stiff enough to keep deflection within the specified tolerance. This will limit the effect of rippling to an acceptable level; it will also provide adequately tight joints to previous concrete below and at the side. It will limit shutter movement during concreting so that there are no ill-effects on half-set concrete. Another point to be dealt with at the design stage is the pattern both of the layout of facing materials or panels and of the ties. By arranging these in a logical fashion, a far better appearance can be achieved. Construction joints, the most conspicuous part of the pattern, are usually laid down by the structural designer, but it may be possible to modify positions in the interest of appearance (Fig. 34). Where grooves in the concrete are specified, every effort should be made to hide the tie marks there and perhaps arrange specially to have grooves for this purpose. The design should be such as to facilitate the construction of accurate formwork joints where the two adjacent faces of the form material are level with one another, preventing the step which is so often seen.

Form surface

The facing of the formwork is most important, because it is the material which is actually in contact with the concrete. Many of the materials used have a potential for absorption: they can absorb air, water, and even to a small extent cement. This is particularly so because during concreting the pressure which is applied by the wet concrete is quite considerable. If any of these normal constituents are extracted from the concrete mix, the concrete colour will change to some extent. This effect may be unimportant if it is totally uniform, but often this is not the case. The effect of the form absorbing air is to reduce or eliminate blowholes, which is normally highly to be desired. Very absorptive materials such as fibreboard thus have attractions; unfortunately, such materials have low durability, and so may be unsuitable even for one use. Conversely, impervious materials such as steel and plastic encourage blowholes. Timber-based materials tend to absorb cement, and this affects the rate of absorbence of water. Therefore, where the appearance is important, materials of different numbers of uses should not be mixed together (Fig. 37).

Timber materials such as plywood show grain. Because of the differing absorbency of sap wood and hard wood, a print of the grain pattern on the concrete is often seen, due to differing absorbency of water from the concrete. If such material continues in use for some time the softer wood compresses, and then the print-out on the concrete is not merely coloured, which will reduce during use, but there is also the physical shape of the board where parts of the grain stand out.

Plywood which has been damaged will normally print out on the formwork. Examples are dents in the plywood, damaged corners,

Fig. 37. The dark board markings are where new boards have been mixed with the originals

or rippling where the plywood has been overstressed and does not recover when the concrete pressure is released. Steel formwork suffers from distortion and dents, and the edges may get damaged.

Foreign matter in the concrete

The concrete can get contaminated from the facing material in a number of ways. The most common is from sugar. In all timber, including plywoods, there is a chemical which when subjected to the sun's rays produces a form of sugar. This reacts with the concrete to produce retardation (the setting of the surface concrete is inhibited). While this is normally so minor that it passes unnoticed, perhaps the merest mark on the concrete at some point, it can in some cases with tropical hardwoods produce retardation of major import. Even in a single batch of boards there may be considerable variation of the effect, and thus it is difficult to eliminate by trial and error. One safeguard recommended is to apply lime to any such panel before use. Timber which has been coated with one of the heavier grades of plastic is effectively immune to this problem, but the lighter grades are not as impervious as might be expected.

Quite a lot of timber has a natural dye in it, and this can leach out, tinting the concrete slightly. Timber-based materials may have rough bits sticking out which can get caught in the concrete. All bits and pieces, dust, shavings, binding wire and so on should be removed from the formwork.

Painted coatings

Many materials are available with a painted coating. It is important that any such paint be compatible with concrete, and normal paints are unsatisfactory for this purpose. It is common to use polyurethane. If this is to be attempted under site conditions, it is most important that dry, reasonably warm, dust-free conditions are found, that the timber is carefully cleaned and that it is oil-free. It is extremely difficult to repaint or repair such formwork after it has been used.

Tape

A self-adhesive plastic tape can be used to seal across a joint. Because of the different absorbency of the adjacent surfaces, it will almost always appear as a slightly different shade of grey on the concrete. Despite its small thickness, the line at each side will be apparent. If it is not effectively stuck down — and this may be difficult because of the oiliness of the surface — it may lift during concreting and become trapped in the concrete itself. Another way of making joints tight is to use foam sealing strip. usually self-adhesive, set between the two faces. It is important that it is put so that the edge is exactly level with the edge of the form, otherwise it will protrude and get involved in the concrete, or form a minor groove to leave a rib on the finished concrete.

Steel formwork

Steel formwork presents the problem of rust. There are various treatments available, and an appropriate release agent should be used as an inhibitor.

The use of high-gloss surfaces produces a smooth, almost glasslike concrete. Unfortunately, this accentuates the number and size of blowholes, and may give rise to black markings on the concrete (Fig. 38).

Smooth surfaces from steel or plastic are more susceptible to crazing. Joints are always difficult to bring up to the same standard as the material to either side, and this is particularly so in the case of high-gloss finishes. Impervious plastic is used to form textured and patterned finishes, but normally the effect of shape over-whelms that of colour, and so this approach can be adopted successfully.

Release agents

The basic purpose of a release agent is to enable the formwork to be removed easily, with little effort and without damage to either the concrete or the formwork. The analogy of greasing a tin when cooking will immediately be seen. All sorts of materials have been used over the decades, but it is now accepted that a limited group of oils will serve the purpose satisfactorily, and that virtually everything else has major disadvantages. Some of these oils are not neat oils, but emulsions. Milk is an example of an emulsion of oil in water, and behaves very much like water. The converse, water in oil, is not capable of being washed away with water or rain. Release agents are categorised into seven groups.

Fig. 38. Black marks from very smooth formwork

Type 1. Neat oil. While this type will give better uniformity of concrete colour, it encourages blowholes; it gives a little oil discoloration and hydration discoloration. It is normally better to use type 2.

Type 2. Neat oil with emulsifier. This also gives uniform colour, but is more successful in getting rid of blowholes. If it is provided too generously it can bring retardation and hydration discoloration problems.

Type 3. Mould cream emulsion. This is on a par with type 2, but is virtually immune to problems of oil discoloration when properly applied. However, an excess will give some slight retardation and hydration discoloration.

Type 4. Water-soluble emulsion. This was once very popular and effectively reduces blowholes. Because severe hydration discoloration and retardation frequently occur, it is virtually obsolete.

Type 5. Chemical release agent. This acts not simply as a lubricant between concrete and form, but actually reacts with the extreme outer surface of the concrete, in effect causing very minor retardation. It is normally applied in a thin layer by spray, partly because the material is expensive and partly to avoid over-application, and is generally satisfactory.

Type 6. Formwork sealers and coatings. Strictly speaking, these are not release agents, but barriers preventing undue absorption of release agent by the form. They may be paint or wax treatments which impregnate the surface. A release agent should be used as well, but should be checked for compatibility.

Type 7. Other. This covers any method not listed above. An example is the use of a silicone treatment. Cost prevents the use of some methods.

All release agents should be applied evenly and sparingly. This may be by brush, rag or spray, the latter giving the thinnest coating. Absorbent surfaces should have a pre-treatment before the first application for actual use, whether it be one from category 6, or a preliminary coat of that chosen for repetitive use, to avoid problems such as that shown in Fig. 33.

All release agents have some effect on concrete colour. For important applications a trial should be conducted. For steel, type 5 should be used to inhibit rust. Types 1 and 4 are seldom available, and should be avoided. Further information is available from the suppliers, and from refs 2 and 7.

Joints

The two important points regarding joints are that the joint should be tight, preventing leakage, and that the pieces of material to either side of the joint should be exactly level so that no step in the concrete is created.

To achieve tightness is difficult, especially if one is trying to prevent the escape even of air. Ideally the mating surfaces should have some resilience, so that when the joint is tightened they will squeeze to one another. This means that timber-based materials are likely to give better results than metal. An alternative approach to sealing is to use tape, or put foam strip in the joint, as described above. Methods of achieving a smooth joint are discussed in Chapter 6.

While the use of a release agent is desirable, it gives dust and dirt a means of sticking to formwork, and it is important that the face material should be clean before concreting starts.

Concreting

The contribution of the formwork to successful concreting is significant. It is important to erect it correctly as intended, and before connecting starts to make sure that it is fully tightened and secured at all points. Movement is often the cause of problems which could easily have been prevented with a little care beforehand. While concreting, care should be taken to remove any temporary spacers or the like, so they do not get concreted in.

Stripping and curing

To achieve constant colour, it is important to have a constant curing regime: stripping should take place at a similar age for all the concrete, and whatever curing is subsequently undertaken should be applied to all parts of the concrete for a similar period. Until the concrete is stripped it is effectively safe from damage. Thus there is an argument for retaining the formwork in position as long as practical; however, this may not be consistent with a standardised treatment. It is always possible to provide protection to the corners of columns and the like which are particularly vulnerable to damage.

Improving the concrete

When the formwork is stripped, it will almost always be necessary to fill the tie holes with mortar. A flat surface may be required, or a feature made by recessing the filling a few millimetres. It is extremely difficult to achieve a result which is not noticeable. If the quality of the concrete face does not meet the specification, it may be possible to adopt some 'making good' measures to improve it. This should only be done if the faults are serious, as making good often makes matters worse, not better, owing to the difficulty of matching the mortar used for the repair to the existing concrete.

Textured and patterned finishes

In many cases the structural designer may feel that a more acceptable finish can be obtained than that produced by plain formwork, and will specify a patterned or textured surface. There is a range from a surface which is effectively not quite smooth, to one in which there are deep indentations. That most frequently seen is produced using unplaned timber. It may be that it shows the saw marks, or alternatively has been weathered naturally or treated artificially to emphasise the grain. This pattern will print out on the concrete (Fig. 39).

There is also a variety of sheets of patterned materials from a few suppliers, all in some type of plastic. Other patterns can be manufactured specially. These can produce interesting effects. Whatever type of texture or pattern is intended, it is particularly important to consider the design of the forms to make sure no unsatisfactory details occur. Both joints between plastic sheets and construction joints must be carefully thought out to ensure that the effect is not spoilt (Fig. 40).

Treated finishes

There are a number of ways in which concrete direct from the form can be treated or modified.

Fig. 39. The impression of timber created by formwork

Fig. 40. Concrete from a plastic form liner, showing darker lines at the construction joints

Acid etching. In this method the immediate laitence of the surface is removed.

Sand blasting. This is normally with some kind of pellet, not sand, and has a similar effect to acid etching (Fig. 32).

Retardation. The discussion above has been about the unwanted effects of accidental retardation, but by applying appropriate retarders it is possible after stripping to expose the aggregate in the concrete.

Tooling. There are many variants which involve mechanical removal of the surface to a lesser or greater depth, providing a rougher surface and exposing aggregate.

Shaped concrete. An example of this is to fix rope to the formwork in such a manner that it will be left behind when the shutter is stripped and can then be pulled from the concrete, exposing some of the aggregate and leaving an interesting pattern.

Fig. 41. 'Elephant House' finish

Another example is the 'Elephant House' finish, in which nibs are created by the form, and then the ends of the nibs are hammered off to expose the aggregate (Fig. 41).

In all these cases the choice of aggregate both large and small contributes to the overall effect. Unfortunately, almost without exception such treatments require better quality formwork, as flaws which appear in the face of the concrete are seldom only superficial. Any attempt to apply the techniques as remedial measures to a poor quality surface will thus result in disappointment.

Avoidable problems

Where it is possible to have an input into the design of the structure the following points should be followed. Firstly, avoid

Fig. 42. Chapel at Ronchamps by Le Corbusier

plain areas of any size. It is extremely difficult to get consistently even colour, which is what people always expect. Provide features at the joints. This means not merely specifying what the feature will be, and designing the reinforcement cover to accord with it, but also deciding where various joints should be, and detailing reinforcement laps to suit. Textured or patterned surfaces can be used, either in a series of panels with appropriate joints, or perhaps as a larger overall scheme. The problems of construction joints must not be ignored.

Given sufficient enthusiasm and money, it is possible to create almost anything. If these are not available, it may be better to accept that concrete is a material with a character of its own, and to make the most of that. If one actually wants to build like glass or marble, pehaps it is better to use those materials. The most famous example of unadorned concrete is the work of Le Corbusier in France. His chapel at Ronchamps is shown in Fig. 42.

6 Practical details

Over the years experience has led to the production of a number of standard approaches to the more common formwork applications. Details have evolved which are straightforward and successful in use. This chapter describes such formwork, made in the main from basic materials, with the addition of some of the simpler pieces of formwork equipment.

Columns

Formwork normally consists of a surface, framing or stiffening, and some form of tying to resist the concrete pressure. The common construction of a column form consists of four pieces, one for each side, which are tied round externally to prevent them opening under the effect of the concrete pressure (Fig. 43). The panel is usually of plywood, with vertical timbers behind it. The ties will consist of traditional metal column clamps in four pieces (see Fig. 27). The important points to consider in such a form are the relationship of the ply thickness and the spacing of the vertical members, and the size of the vertical members in relation to the clamps and their spacing. The details at the corners where adjacent panels meet are also important, because they are fundamental in ensuring that a column is of the correct size and that leakage is controlled. The ply face of one pair of sides should be the exact size of the column, and the other pair should be wider by twice the thickness of the plywood. The vertical edge timbers should oversail the narrower sides, thus forming recesses into which the other sides can fit neatly and accurately (Fig. 44).

Where a column has to have a splayed or chamfered corner, it is tempting to provide a timber fillet, but this is difficult to fix efficiently; it is much better to use a larger piece of timber from whose corner the shape of the fillet has been removed. This is set

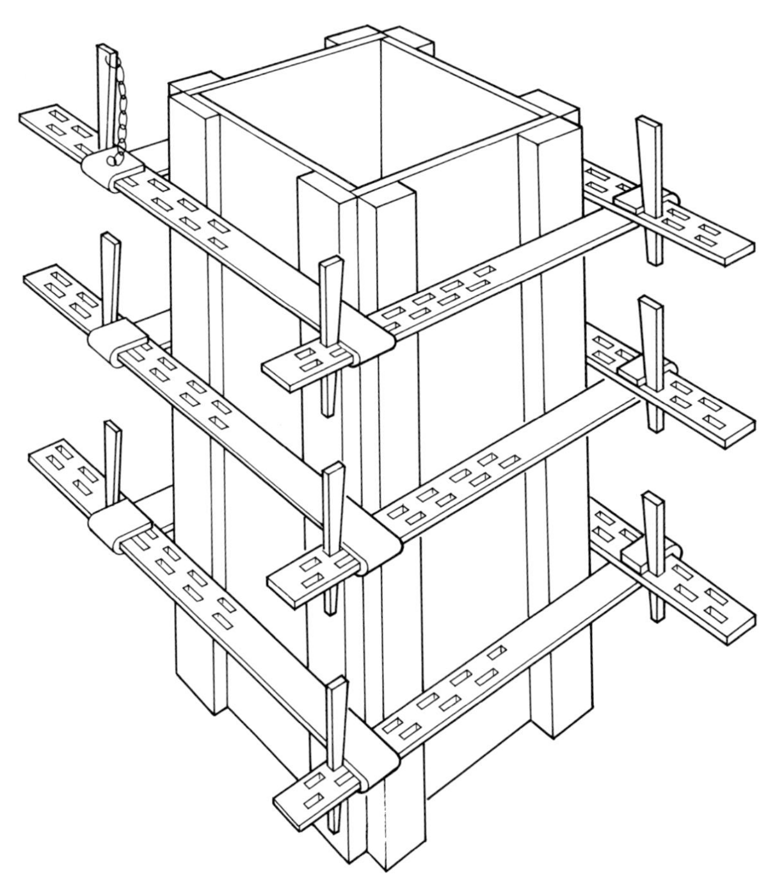

Fig. 43. Typical ply and timber column form

into the framework of the face panel so that the edge of the splay is level with the plywood (Fig. 45). The traditional column clamp will normally be held up by a few nails fixed into the side of the panel. Clamps are of limited strength, and should not be used for columns with sides larger than 750 mm.

This approach to column formwork is suitable for rectangular cross-sections, but it can also be used to produce other shapes, by the addition of formers within a surrounding rectangle. This can be adopted for circular columns too, but it is also possible to adopt the classical solution of a circular tie around a circular form. In hoop tension, a minimum of material is needed and the effect of

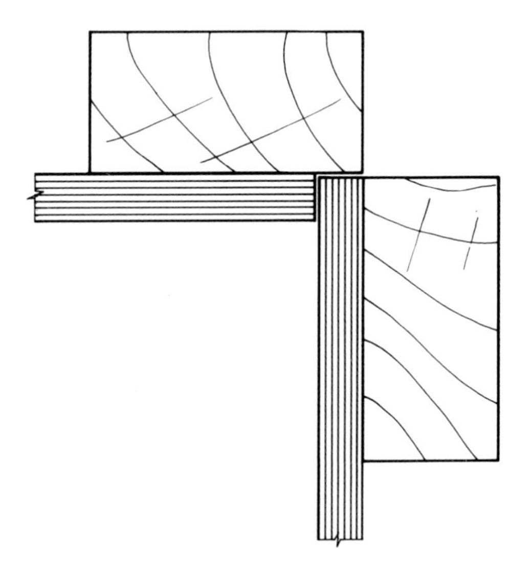

Fig. 44. Detail of a corner

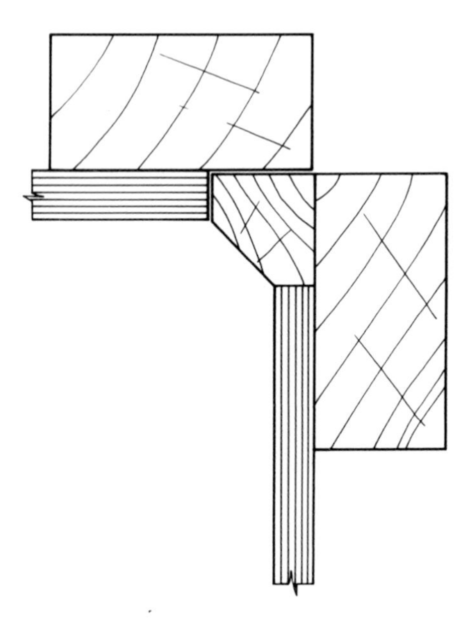

Fig. 45. Detail of a corner with a splay

pressure is to improve the circularity rather than to convert it to an oval. This principle is applied where pipes or cardboard tubes are used. Purpose-made GRP formwork follows the same principle for circular columns.

The column form will sit on a base of some sort, and it is necessary to seal the bottom to minimise leakage of concrete. The most satisfactory approach is for opposing faces of the form to be pulled tightly to a concrete upstand. For a column this would mean having a piece of concrete below of the same cross-section as the column to be cast. This small sample piece is called a kicker. It is important that it and the shutter which will nip on to it are the same size, to a high degree of accuracy. Such a kicker should have a level top surface, and the form should overlap it by a small amount. If the surface of the kicker is not true, or if there has been minor damage to it, the level line around the kicker will be destroyed, and the line of contact of new concrete may be contained in a zone of 20 mm or more in depth. The length of overlap of the column formwork onto the concrete must therefore be appreciably greater than this to provide a seal all round; a figure of 50 mm is sensible. A greater length could mean a higher kicker, and the quality of the joint would fall.

The formwork for the kicker must be held accurately in position, as it dictates the bottom position of the column. If it is to be cast with the concrete below it, perhaps a slab or base, it is not simple to position it correctly without any risk of movement. Where there is heavy reinforcement this provides something against which to anchor it. If on the other hand it is built a day or two later, it is considerably easier to set the formwork out accurately and, by nailing to the concrete, hold it in position. However, this takes no account of the problems of compaction of concrete, and it is more likely that the monolithic construction will lead to better concrete quality. The height of the kicker should allow for the lap length and another 25 mm or so to allow for minor errors of levelling.

Walls

A column which has one dimension so great that it is impractical to tie around it can be considered as a wall. While there are many features in common, the important difference is that tying has to be done along the length of the formwork, typically by ties through the wall. Construction is like that for a column, in that there is a surface, framing and a tying system. However, the framing system usually comprises two layers, one immediately behind the facing material and a second heavier framing member behind that, through which the ties are put (Fig. 29).

The two framing members will be at right angles to one another. If the member closest to the concrete is put vertically it is described as a stud, and the horizontal member behind it is called a wale. The controlling factor in the design of sheeting in this arrangement will be at the bottom where the pressure is greatest. Whatever spacing is necessary for the facing supports will thus apply to the full height of the formwork. The waling will help to make the wall straight in its length, but any vertical straightness is dependent on the studs close to the face, which will normally be of a lighter material size. The alternative arrangement, which is more popular, is to have horizontal members immediately behind the facing (also referred to as wales), and behind them to have substantial vertical members called soldiers. The spacing of the wales can correspond to the concrete pressure, being wider at the top. In a domestic storey height it is possible to tie such soldiers at the bottom just above the floor, and over the top of the concrete, but the design can be applied to much higher walls with ties going through the concrete (Fig. 46).

Formwork can be built up for each use, in a form suitable for a particular configuration. The maximum size of piece which can be handled is dependent on its weight if it is to be moved by men. A limit would be about 40 kg. It is always possible to move larger pieces with several men, but care should be taken to see that they do not strain themselves lifting it. If the pieces are to be crane handled, the likely limitation will be the effect of the wind on large panels.

For formwork which will only be used once and then be dismantled, the minimum of nailing to provide a stable and proof form should be provided. If it is to be rehandled elsewhere, greater care should be taken; failure to do this will result in the framing coming adrift from the facing, time lost in maintenance of the forms, and possibly the necessity for remedial work to the concrete where the quality is inadequate. For wall forms, both panel layout and tie arrangements require careful thought. The panels should be so arranged that they are of a suitable size for

Fig. 46. Formwork with steel wales and soldiers

reuse in a number of applications; in general, ties should be used at the same point in each panel each time, and the repetition of use will apply to the panels too. Where ties cannot be used through the same hole it is necessary to plug them, which is a time-consuming operation and, if the concrete is of high quality, unlikely to produce an acceptable finish.

It is usual to have one line of ties close to the construction joint, so that the nip-on of the new form will be as tight as practical to the earlier concrete. It is convenient to design the form so that the top tie will be just over the top of the concrete to be poured, enabling a repetition as the form goes up the building. Alternatively, to put a tie in the concrete close to the top enables the panel to be held tightly to the top edge of the concrete, provided the height of nip-on is limited. It also means that a similar tie will be used just below the top of the new lift, and by replacing the bolt in the tie hole a support can be provided for the form to sit on while it is erected above the present position (Fig. 47).

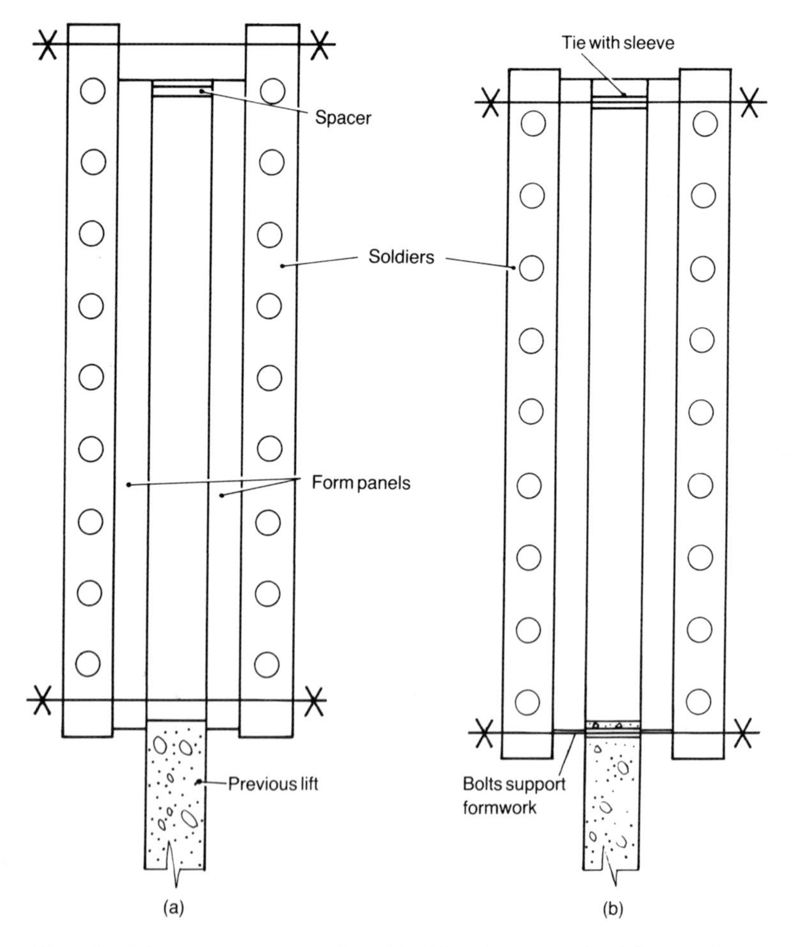

Fig. 47. Tie arrangements for climbing formwork: (a) ties above top of lift; (b) ties through top of lift

Ties should be used without spacers, as it is impossible to prevent their being overtightened. It is important to do all ties up tight to limit take-up between different parts of the formwork when the load is applied. For forms for which considerable use is anticipated and for which particular care will be taken of the face to prevent it being rejected for reasons of damage, it may be appropriate to use both glue and screws to ensure effective joints in the formwork.

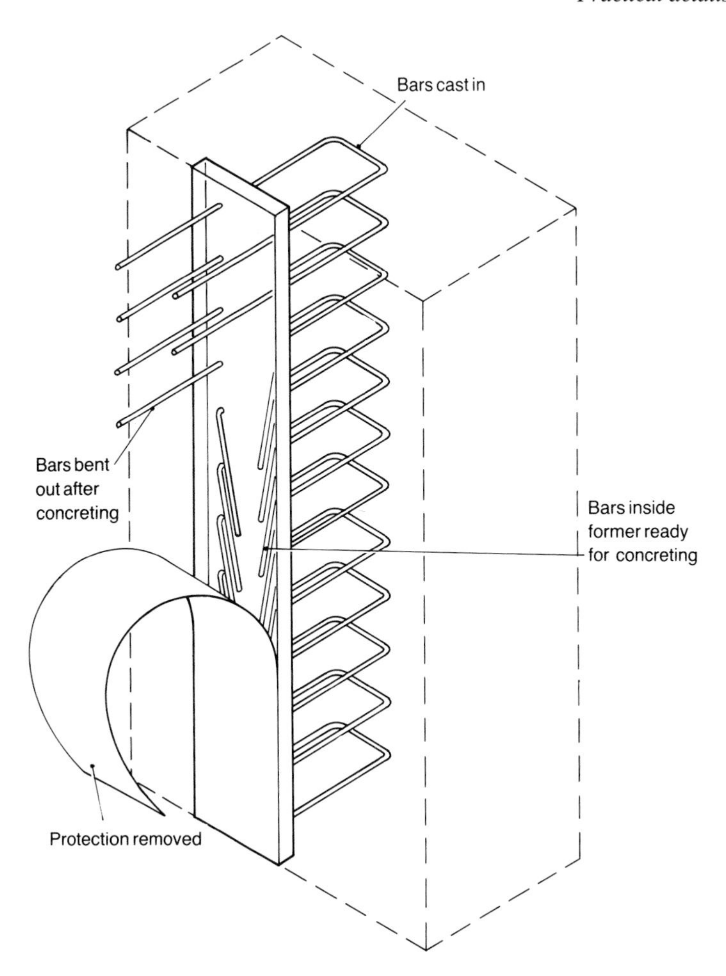

Fig. 48. Construction joint former

Adjacent panels require connecting, and nuts and bolts are frequently used. These enable two adjacent panels to be drawn strongly to each other, making a tight joint, but will not ensure that the two contact surfaces are in line, due to tolerances in the bolt holes.

It is usual to make the panels with an accurate thickness, so that when a framing member is used to line up the back of the panels, the front will also align.

Construction joints

It is seldom possible to construct a complete unit of concrete in one pour, and so a construction joint has to be used. This results in reinforcing bars passing through the face of the form. There are a number of ways of doing this. It is difficult to remove formwork which has been drilled and bars made to project. Instead, frequently the form is split so that a notched section meets a straight one, making a fairly tight joint. Another approach is to use expanded metal, the small loss of grout normally being acceptable. Smaller bars may be temporarily bent back within the outline of the concrete and wrapped to ensure they can easily be bent out after removing the formwork panel. There are a number of proprietary designs for this; an example is shown in Fig. 48.

Beams

Most beams support slabs, and the first thing to decide is whether the beam must be constructed on its own or whether it will be better to construct the slab with it. Unless the beam is of considerable size, the quantity of concrete involved will be small, and unimportant in relation to that needed for the floor. Another aspect is access. It is necessary to provide adequate working space to construct the beam on its own. For beam and slab construction together, the soffit form will provide access for the concreting, and less access will be needed where the two are constructed together.

Beam formwork consists of a bottom or soffit form and two sides. While plywood is a first choice for many items, it may be more sensible to use timber for the soffit of a beam. Provided the width of the beam is within that of the available timber, a single long piece of material for the soffit is attractive. It will be thicker than plywood, and so the number of supports needed to keep the deflection within the limits will be fewer. The beam sides can sit on this soffit form, or alternatively be clamped to the sides and sit on the joists below. To ensure the correct dimension, the latter arrangement is to be preferred, although if there are various widths to be produced setting the forms on top may be attractive. If plywood is chosen for the soffit, it will frequently be made into a panel. The edges and perhaps an intermediate member will be of timber capable of spanning some distance, reducing the need for cross-supports. These supports will rest either directly on a vertical support such as a pair of props or scaffolding, or in some cases it may be sensible to have a pair of heavier bearer timbers, supported on verticals at greater centres. Fig. 49 shows a typical arrangement.

Stability of formwork up in the air is always important, but if a beam is to be constructed on its own, or if the support for the slab constructed as the same time is with telescopic centres or the like, not requiring direct support from below, then the stability of the whole arrangement is particularly critical. In the length of the beam it should be fairly simple to stabilise it against the columns or walls which support the ends of the beam. In the other direction, it will almost certainly be necessary to provide diagonal bracing to ensure stability.

The sides must be held in a vertical position to resist the lateral pressure of the concrete; the bottom of the beam side must be held

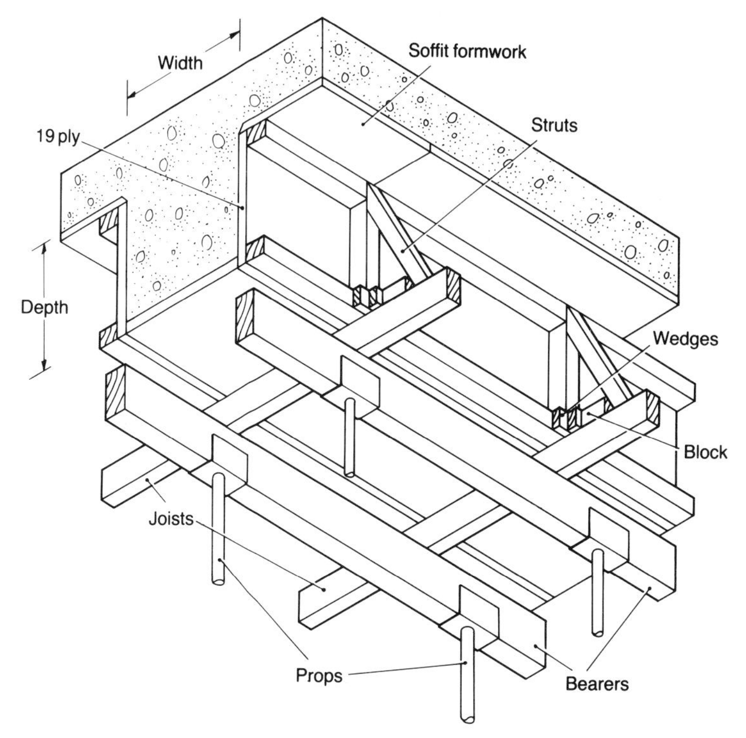

Fig. 49. Typical beam formwork

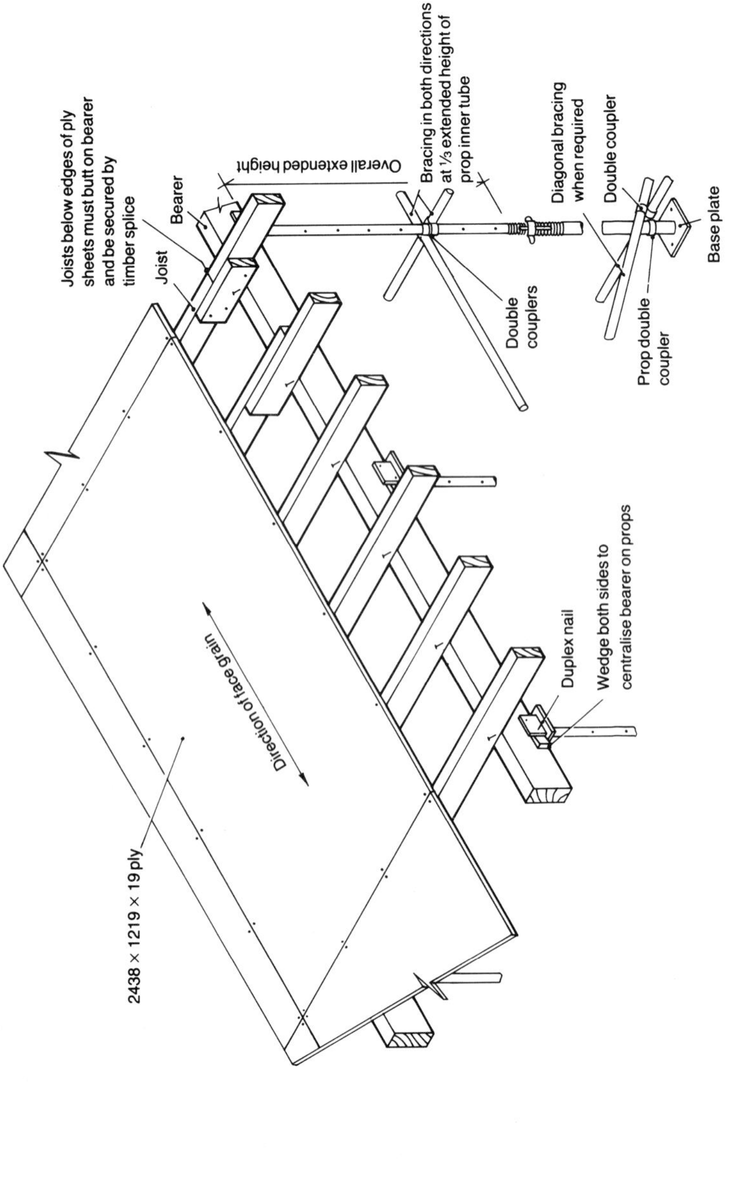

Fig. 50. Typical floor formwork

tight to the soffit, so that leakage is effectively controlled. In addition, the beam side must be held at the correct height, so that the top is in the correct position.

To hold the sides in vertically, it is common to use pieces of timber on the rake, nailed top and bottom, which act as struts. To hold the bottom in, the shutter may be held tightly while it is nailed down, or alternatively blocks or a runner be placed on the joists, leaving space for folding wedges to tighten the bottom of the beam side against the soffit. Proprietary clamps or small push-pull props can be used to achieve the same purpose. If a beam is of such a depth that using a raking support does not achieve adequate support to the sides, it will be necessary to treat it as if it were a wall, and tie at a level above the beam soffit.

Soffit formwork

The classical solution to this problem is exactly the same as that for a floor in permanent construction. The surface consists of a facing, such as ply; secondary beams are placed beneath this, the spacing being arranged to limit the deflection of the facing. These secondary beams span onto primary beams, analogous to walls in a house; however, these primary beams will normally be supported by props, or a scaffold of some sort (Fig. 50). The concept is straightforward, and the mathematics fairly simple. It is practical to use tabular data. Potential problems arise principally in erection, but also in stability and use. Where the storey height is no more than 2·5 m the erection of props, using a lacing tube fixed at shoulder height, enables a stable structure to be created without difficulty. It is then fairly simple to place timbers on top of this arrangement, as it is within reach of the ground. For taller storey heights there are problems of access, and a scaffold of some sort is appropriate. In all cases it is important to establish proper stability, so that there is no risk of the whole structure falling over during erection or later. Where the floor is within a room little difficulty will be found, but where it is in an open space it will be necessary to use diagonal bracing, or tie it to columns.

Because it is normally impractical to use a single piece of facing material, joints must be considered. In the direction of greater strength of the board or sheets, spanning across the secondary members, it is essential to have the end-to-end joints on a secondary member. This will dictate the layout. Along the sides of the sheet of material it is normal to butt the boards, and if this is done tightly and the board nailed down, leakage will be minimal. If the quality so produced is inadequate in the context, it may be necessary to create an extra framework above the main support system so that light timbers can be used to overlap the joint in this direction, as well as underneath the end joints. It will be seen that any joints in the secondary beams have to be in line, end to end, if they are supporting a joint in the sheeting material. However, other secondary members can be lapped over the primary member. For primary members it will almost always be necessary to butt them in the forkhead or above the prop.

It is frequently necessary to arrange for a camber in a slab or beam. The form has to be set up above the theoretical line to allow for

(a) aesthetic considerations. A flat soffit appears to bulge downwards, and so a camber is put in to make it look right.
(b) deflection of the concrete. When the formwork is removed, the concrete will deflect, and it is possible to make an estimate of this deflection.
(c) deflection of the temporary works. Depending on the structural arrangement, there may be a significant deflection when the formwork and falsework are loaded, and allowance must also be made for this.

The specification should be studied to establish if there are any special requirements.

Construction joints have to be formed in soffits. This may present problems, because the formwork on the loaded side will deflect initially, and the concrete added subsequently on the previously unloaded side will cause the formwork to descend further. This has the effect of loosening it from the earlier concrete, and grout frequently runs in below, spoiling the effect. The position of the construction joints and the support arrangements should be considered together.

Telescopic centres

The formwork and falsework erected to construct a beam is seldom fully loaded; the spare capacity can be used to support the soffit formwork for the slab which the beam itself will ultimately

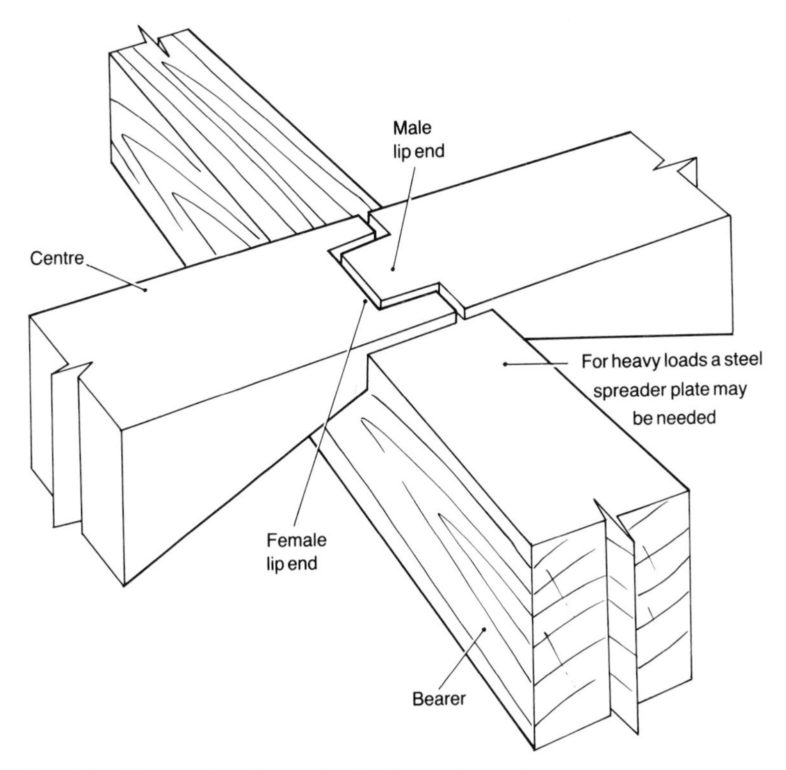

Fig. 51. Supporting the ends of telescopic centres

support. Any type of beam can be used to span from one beam support to the next to carry the soffit, but telescopic centres are especially suitable for this purpose (Fig. 24). The two pieces, telescoping one inside the other, can be adjusted so that the tongue at each end fits on an appropriate seating on the beam sides (Fig. 51). One of the advantages of this arrangement is that access to the space underneath is unimpaired, so that it can be used for a variety of purposes.

Although this is a useful set-up, experience shows that there is a higher than average risk level and so particular care should be taken. The actual tongues on the end are relatively small, and must be carefully positioned. For some larger centres, when fully loaded the bearing pressures become excessively high, and it may be necessary to provide a small steel plate to spread the load if the support is from timber. Most centres are provided with a locking device which will prevent any futher movement of the two halves, but any shortening which occurs, for example through concrete vibration, could result in the ends slipping off their suppport. It is important to provide or check the load path down to the main support system. If a ply and timber panel is used for the side shutter of the beam, it will almost always be necessary to add additional vertical timbers to carry the load from the top frame to the bottom frame. From there on checks should be made that there are no overloads occurring. Telescopic centres are provided with a preformed camber, which should approximately flatten out under full load.

Splays

Because of the stress concentration at corners, it is frequently desirable to use a splay to ameliorate this condition. If this is small, it may be possible to add a fillet to the top of the beam side, but if it is larger, it will be necessary to make special panels, and provide a support system for them. An example is shown in Fig. 52.

Curved formwork

A curved surface can be formed using sheet material pulled to the appropriate radius. It is always difficult to bend correctly the extreme ends of such a panel, and to hold them in the correct shape. An alternative approach which may be appropriate in some cases is to use a series of flats with a small angle at each joint. For a circular column shutter such slats could be as narrow as 20 mm, while for a large tank a series of panels 300–600 mm wide may produce an acceptable approximation to a circle. The offset should be calculated, to ensure that proper cover to the reinforcement is being provided at the centre and ends of the panel on the two sides. The manufacturers provide data on their plywood, setting out to how tight a radius it can be bent. Normally these figures are optimistic, and not really practicable to achieve on site; an increase of 25–50% is sensible. Plywood can be thinner than it would be in a straight wall, because curving it has the effect of providing additional stiffness. For very sharp curves, 3 mm ply or hardboard may be able to give the smooth curve required.

Where a band saw is available, it is possible to cut wales to the

Fig. 52. Splay formwork to a bridge

required curve. This enables a shutter panel to be made with horizontal framing. If these members are of any length, the loss of strength due to that which has to be cut away may be significant, and the design should be based on the minimum cross-section. An alternative approach is to use studs on the facing, to pull the panel to the required shape, and to use a waling with packers behind to hold it to that shape. The wale approach produces a more satisfactory arrangement, but involves more work in the first instance.

Proprietary equipment is described in chapter 3.

Top formwork

As a liquid, concrete, particularly when vibrated, tends to find its own level. If a sloping surface is required, therefore, it is likely to need some formwork to the top. Depending on the concrete mix, the depth of section, the amount of reinforcement and the accuracy required, it may be possible to omit top formwork, perhaps up to 30° or so. Top formwork always causes bubbles in the surface, and some small holes will help the air to escape. Where formwork is provided it must be designed for the full concrete pressure.

Plumbing

Wall formwork is intended to create concrete which has a correct position and orientation. Formwork panels may not be true, and panels can move if not well anchored. Frequently the most practical method of plumbing a wall form and enabling it to withstand the wind is to use raking props at both sides. They can be placed as a single row or, if the wall is tall or tending to be distorted, additional props can be used to establish a satisfactory

Fig. 53. Lining of vertical formwork achieved by adjusting the tail of a soldier

arrangement. The effect of raking props and the vibration of the concrete being placed causes a tendency for the shutter to rise. Provided effective nipping on to a kicker has been achieved, all should be well; otherwise, vertical anchorage of the shutter itself should be provided. For larger forms, where standard telescopic props would be too flimsy, it is possible to use heavy duty prop units, or soldiers which have end fittings to turn them into push-pull props. Such an arrangement also has the advantage that when anchored, such props need only be put at one side, because they will act both as a strut and a tie. For really large formwork wire ropes can be used on both sides with turn buckles or Tirfors to tighten them.

When formwork is being used to extend a wall upwards without an intermediate slab, the soldier and wale design can have its own straightening facility built in by using soldiers which extend some way below the formwork panel itself. The bottom is wedged or jacked from the wall to position it accurately in a vertical position. An example is shown in Fig. 53.

It is normally necessary to plumb columns in the two main directions, but otherwise the remarks above are applicable.

7 Stripping

It is necessary to postpone stripping until the concrete is strong enough to resist the damage which can occur on or after exposure. Early stripping can overstress the concrete, causing cracking or worse, or on a more minor scale the surface may be damaged. It should also be considered that formwork helps to protect concrete against frost.

For a particular concrete mix, a reference graph setting out the rate of gain of strength can be established experimentally. This assumes a temperature of 20°. Variations from this which actually occur can be recorded, and account be taken of them to establish from the graph when the strength is adequate. The hotter it is, the more rapid the gain in strength. If the concrete temperature falls too low, all hardening ceases and the period before stripping can be greatly extended. It is thus wise to apply some form of heat in cold weather, for example using hot water for the concrete, to ensure that stripping can take place after a reasonable time. As well as providing heat, it is sensible to retain the heat created by the concrete itself. Steel formwork is no protection at all; plywood is better, and it is possible to add plastic foam to a form if there is likely to be protracted cold. Open surfaces where there is no formwork should also be protected with some form of insulation, not only against the cold but against the drying effect of the wind and possible overheating due to the sun.

The strength required at stripping will depend on the loads anticipated. For a soffit this will be the self-weight plus at least a working load; for walls it will be the wind. As well as this basic strength requirement there is the need to strip cleanly and have corners which are not readily broken. In summer conditions this may be as soon as next day, but in cold weather it can be considerably longer. It is necessary to achieve a concrete strength of at least 2 kN/mm^2 as a protection against frost damage, and some researchers recommend a higher figure.

The concrete strength can be established in a number of ways. There are tests which depend on pulling out an item cast into the concrete, or breaking off a small piece partly separated from it. From the results of these tests the strength can be inferred from laboratory tests on the same mix done as a calibration. A Schmidt rebound hammer can be used on an exposed surface; calibration is required for this as well. Gamma ray and ultrasonic testing have also been tried.

The cube strength is not very helpful, because the rate of gain of strength of a cube in a cast iron mould differs from that of the same concrete in a different situation. If the cube can be put in a hole in the slab at a point where the temperature will be substantially the same, it may prove a reasonable guide. Alternatively, a temperature-matched curing bath can be used. A temperature probe is placed in a part of the concrete which is expected to be relatively cool, and the temperature of the water in the bath is electronically matched to this temperature, thus curing the cube at the appropriate rate.

Another method involves recording the temperature of the concrete and of the air at least twice per day, or better still producing a continuous record. These figures can be used to establish the strength from a reference graph.

Where no calibration is available, the use of the CIRIA formwork striking tables[8] will provide answers. Data needed include the temperature of the concrete at placing and subsequent air temperatures, cement type, concrete grade, characteristic cube strength, minimum dimension and formwork material.

The job specification frequently gives guidance on stripping times, but these are normally conservative figures which are simple to use but unnecessarily lengthy.

Propping and re-propping

It is usually possible to transmit the load from the fresh concrete of a new slab to the area directly underneath. At the start this is likely to be the ground, and there will seldom be any shortage of load capacity. Similarly, further up the building one slab may well be able to support another. With rapid construction, however, the strength of the lower slab may be too limited, and other arrangements have to be made.

Where a propping system goes down to a solid slab on the ground, all the load will be taken down it. Any slabs above the ground will only take load if they are allowed to deflect. It can thus be seen that the lowest level of props will be carrying the total weight of all the slabs above.

If a fresh slab is supported by another pair below, loads do not always divide equally between them. The lowest slab may have to carry over 60% of the weight of the two slabs above, depending on the sequence of the loading and stripping. For further information on this subject see Appendix M of the Code of Practice for Falsework.[3]

Maintenance

When the formwork is struck, it should be cleaned and inspected for damage (see chapter 8). For timber and plywood formwork, repair may consist of fixing pieces which are coming apart by re-nailing, perhaps replacing a damaged member, or replacing the facing material.

If it is desired to repair the face sheeting, this can be done with a plastic filler, or in more major cases by infilling with another piece of plywood—a job for a good joiner. The use of a filler will only be successful if the timber is cut back far enough such that the mould oil has not penetrated; otherwise it would be impossible for the filler to get a good hold. Such a filler will almost invariably show on the concrete, owing to its different absorbency.

Formwork of timber and ply can normally be dismantled, and if the state of the materials is good enough, they can be used again. It is most undesirable to leave timber with nails projecting. If these timbers are to be burnt, the nails should be bent over on dismantling; if not, they should be removed. The more thoroughly formwork has been constructed, the more difficult it will be to get it apart for further use.

If the equipment is of steel, site maintenance will be much less practicable. The forms can be cleaned, but it is not easy to remove dents, and the welding of broken parts will prove difficult. They should be returned to a metalworking shop which has adequate facilities. With proprietary equipment in particular it is difficult to judge the grade of steel, and welding techniques must relate to the original steel used to be successful.

Renovation of aluminium alloy members is even more difficult, and apart from straightening minor bends at corners, it is impracticable to do anything on site.

Much plastic formwork is unmendable on site, and it is really only GRP which can be effectively repaired. This can be done by using car body techniques, but the comment about getting rid of the mould oil first is equally important here.

8 Quality

As with all materials, it is difficult to produce concrete which can be described as without blemish. While it is often barely perceived by structural designers, appropriate detailing of the structure can reduce the impact of problems. It may be possible to introduce minor changes on site, for example to break up plain areas with small recesses. By providing strong lines, which should tally with formwork sheeting sizes, the eye is led away from slightly unevenly coloured panels of grey concrete.

Concrete may also be spoilt by weathering, but changes to the outline to reduce this problem will result in more than minor changes to the formwork.

Checking

To ensure that the resulting concrete is up to the required standard, it is necessary to make sure that an adequate formwork design is adopted and all the materials and workmanship are acceptable. Will the intentions be translated into a satisfactory reality? Making sure nothing is going wrong should take place at each stage. A checklist covering formwork items is given in the Appendix; for falsework, see the relevant Works Construction Guide.[2]

Design

On completion, the design should be checked for

(a) concept
(b) conformity with specification
(c) conformity with expected site conditions (including rate of concrete placing)
(d) the mathematics
(e) appropriate detailing.

Even if the design has been prepared from standard data, a mathematical check can and should be done, if a judgement based on experience suggests it is not adequate.

Materials and equipment

Materials and equipment should be checked on arrival to make sure they are what is required by the designer. They should be examined for reduction of section or deterioration, particularly if they have been previously used. It should be ensured that components, such as pins, are to the manufacturer's specification.

Assembly

All materials and items of equipment should be checked to ensure that they are in the intended place, and of the proper type. Material may be of the wrong grade or incorrect cross-section. Dimensions should be checked, to ensure that the form is in the right place and spacings are not too great. It is important that all connections are done up tightly, so that any movement is minimal and leakage is kept under control.

Access

Good access will speed up the work, and improve the quality. For small walls, the floor or ground adjacent will serve; taller walls will need access, probably bracketed from the form itself. At the edge of soffits an unencumbered space must be provided for access, both for formwork and concreting operations. Guardrails, toe boards and ladders will be needed for all these applications. Brick guards will make it safer for those below.

Modifications

Where the exact items originally specified are not available it will normally be possible to accept alternatives, but these must be considered in the context of the original design. Other changes may have to be made as well, to achieve the original object. All must be checked with the original design criteria.

Appendix. Formwork checklist

A checklist for formwork should contain the following items, along with columns in which to show that an item has been checked, what faults there are to be put right, and that a final check has been made. For falsework foundations, tube and fittings, telescopic props, telescopic centres, aluminium components and proprietary equipment see ref. 1.

1. General

Design
- concept
- relation to expected concreting rate
- strength
- stiffness
- dimensions tally with structural outline

Material
- type/grade
- dimensions
- condition

Access
- adequate space
- guard rails
- toe boards
- brick guards
- ladders

Assembly
- setting out
- width
- breadth
- height
- plumb and stability
- inserts fixed
- ties tightened
- all other fixings secure
- sealed against kicker
- panel joints tight

Before concreting
- release agent applied
- formwork cleaned out

During concreting
- removal of form spacers
- watch for movement

Stripping
- nails bent over or removed
- reusable material separated

2. Timber

Material
- size
- species
- grade
- condition
- damage

Joints
- position
- continuity
- bearing between members
- adequate area
- eccentricity
- tapered packs
- wane

Lateral stability
- narrow timber
- timber at an angle

3. Plywood

Type
- make, brand
- timber species
- lay-up
- thickness

Face grain direction

Damage

Repairs
- filling over nail holes

4. Ties

Type
- make, model
- diameter/strength
- water bar

Location

Washer
- size
- thickness

Bolt
- thread condition

References

1. Wilshere C. J. *Falsework*. ICE works construction guide, Thomas Telford, London, 1983.
2. Concrete Society. *Formwork — a guide to good practice*. Concrete Society, London, 1986.
3. British Standards Institution. *Code of practice for falsework*. British Standards Institution, London, 1982, BS 5975.
4. Clear C. A. and Harrison T. A. *Concrete pressure on formwork*. CIRIA, London, 1985, Report 108.
5. British Standards Institution. *Code of basic data for the design of buildings, CP3 Chapter V. Part 2: Wind loads*. British Standards Institution, London, 1972.
6. Health and Safety Executive. *Safety in falsework for in-situ beams and slabs*. HMSO, London, 1987, HS(G)32.
7. Monks W. *Visual concrete design and production. Appearance matters no. 1*. Cement and Concrete Association, London, 1980.
8. Harrison T. A. *Formwork striking times — methods of assessment*. CIRIA, London, 1987, Report 73, 2nd edn.

Bibliography

BUILDING ADVISORY SERVICE. *Construction safety.* National Federation of Building Trades Employers, London, 1982.

GAGE M. *Guide to exposed concrete finishes.* Architectural Press, Cement and Concrete Association, London, 1970.

GLASSFIBRE REINFORCED CONCRETE ASSOCIATION. *Permanent Formwork.* GRCA, Shropshire, 1985. Handbook No. 1

GROCOTT P. C. *Concrete release agents.* BP Aquaseal, Kingsnorth, 1981, 3rd edn.

HARRISON T. A. *Permanent GRC soffit formwork for bridges.* Cement and Concrete Association, London, 1986, Interim Technical Note 9.

HARRISON T. A. *Formwork striking times — methods of assessment.* CIRIA, London, 1987, Report 73, 2nd edn.

HURD M. K. (ed.) *Formwork for concrete — SP4.* American Concrete Institute, Michigan, 1981, 4th edn.

IRWIN A. W. and SIBBALD W. I. *Falsework. A handbook of design and practice.* Granada, London, 1983.

MONKS W. *The control of blemishes in concrete. Appearance matters no. 3.* Cement and Concrete Association, London, 1981.

MONKS W. E. *Textured and profiled concrete finishes. Appearance matters no. 7.* Cement and Concrete Association, London, 1986.

MONKS W. E. *Exposed aggregate finishes. Appearance matters, no. 8.* Cement and Concrete Association, London, 1985.

MONKS W. E. *Tooled concrete finishes. Appearance matters, no. 9.* Cement and Concrete Association, London, 1985.

MURPHY W. E. *The influence of concrete mix proportions and type of form face on the appearance of concrete.* Cement and Concrete Association, London, 1967.

Formwork

National Association of Formwork Contractors. *Technical information sheets (5)*. NAFC, London.

Peurifoy R. L. *Formwork for concrete structures*. McGraw-Hill, New York, 1976, 2nd edn.

Richardson J. G. *Formwork construction and practice*. Viewpoint, London, 1977.

Richardson J. G. *Formwork notebook*. Viewpoint, London, 1982, 2nd edn.

Ward F. *Striated finish for insitu concrete using timber formwork*. Cement and Concrete Association, London, 1972.